Lecture Notes in Physics

For information about Vols. 1–90, please contact your bookseller or Springer-Verlag.

Lecture Notes in Physics

172

Ionic Liquids, Molten Salts and Polyelectrolytes

Proceedings of the International Conference
Held in Berlin (West), June 22–25, 1982

Edited by K.-H. Bennemann, F. Brouers, and D. Quitmann

Springer-Verlag
Berlin Heidelberg GmbH 1982

Editors
K.-H. Bennemann
F. Brouers
D. Quitmann
Freie Universität Berlin, Fachbereich Physik
Arnimallee 14, 1000 Berlin 33

ISBN 978-3-540-11952-4 ISBN 978-3-540-39540-9 (eBook)
DOI 10.1007/978-3-540-39540-9

Originally published by Springer-Verlag Berlin Heidelberg New York in 1982

2153/3140-543210

PREFACE

This book presents a series of up-to-date review articles and contributions on recent research in the areas of ionic liquids, molten salts and polyelectrolytes.

Electron-charge transfer and its effect on various structural and thermodynamical properties and phase transitions in various ionic systems are discussed in detail.

A review is given of the use of molten salt electrolytes in batteries, and many experimental and theoretical results on the static and dynamical properties of polyelectrolytes are reported. The significance of polyelectrolytes in biology is also discussed.

It becomes apparent that similarities exist among the physical ideas underlying the structure and the thermodynamical and dynamical behaviour of the different systems treated in this book.

Berlin, September 1982

K.-H. Bennemann
F. Brouers
D. Quitmann

TABLE OF CONTENTS

INTRODUCTORY ADDRESS

N.F. Mott
Cavendish Laboratory, Cambridge, U.K.

Abstract

A discussion is given of recent work both experimental and theoretical on the existence of a minimum metallic conductivity. In doped semiconductors a sharp distinction must be made between the behaviour of compensated and uncompensated materials, the concept in its original form being valid only for the former. At high temperatures, and thus in liquids, arguments are however given which suggest it retains its validity.

1. A minimum metallic conductivity?

Many fluids show some form of metal-insulator transition, either with a change of specific volume (Hg, Cs), a change of concentration (NaCl + Na) and metal-ammonia solutions, or with variation of temperature as in fluid Se-Te. The interpretation of these phenomena, and of conduction and thermopower in these materials, has involved the concept of a minimum metallic conductivity. This concept has been greatly affected by recent work, both theoretical and experimental, and I think the most useful contribution I can make to this meeting is to survey the present position, particularly as it applies to liquids.

To derive a minimum metallic conductivity, in my original work[1] I started from the tight binding model as it was used in 1958 by Anderson[2], with band-width B and disorder parameter V. Localization throughout the band is thought[3] to occur when $V/B = 2$, and for somewhat smaller values states are localized up to an energy E_c (the mobility edge). For a degenerate electron gas with Fermi energy E_F near E_c but in the non-localized range, the conductivity can be calculated from the Kubo-Greenwood formula [4, 5, 6], and is under certain assumptions to be discussed below, given by[6]

$$\sigma = (\pi e^2/\hbar a)\, z a^6\, I^2 \left\{ N(E_F) \right\}^2 . \qquad (1)$$

Here a is the distance between wells, z is a co-ordination number taken to be 6 for a random array of centres, I the transfer integral and N(E) the density of states. Writing B as the band width without disorder, $B = 2zI$ and

$$N(E_F) = 1/a^3 (V^2 + B^2)^{\frac{1}{2}} , \qquad (2)$$

we find

$$\sigma = [\pi e^2/(4\hbar a z)]\, B^2/(V^2 + B^2). \tag{3}$$

The other method, even less rigorous but sometimes convenient, is to start from the Ioffe-Regel conductivity

$$\sigma = \frac{1}{3}\, e^2/\hbar a$$

which one expects when the mean free path l is of the order a. It is then easy to see, again from the term $(N(E))^2$ in (1), that for $V > B$

$$\sigma = \frac{1}{3}\,(e^2/\hbar a)\; g^2 \tag{4}$$

where g is the reduction in N(E) due to disorder. Mott and Davis[6] take, in the absence of disorder,

$$N(E) = 1.75/a^3 B,$$

the 1.75 arising because N(E) is peaked in mid-band; with disorder it is given by (2). Thus

$$\sigma = \frac{1}{3\,(1.75)^2}\;\frac{e^2}{\hbar a}\;\frac{B^2}{V^2 + B^2} \tag{5}$$

The numerical factors in (3) and (5) are 0.125 and 0.11, near enough together in view of the very crude approximations involved.

The minimum metallic conductivity is defined as the value of the conductivity when the Fermi energy E_F lies at the mobility edge, and if this is near mid-gap we may set $V/B = 2$, giving from equation (3)

$$\sigma_{min} = 0.025\; e^2/\hbar a \tag{6}$$

If E_F does not lie near mid gap, then according to the present author we should write a_E instead of a where a_E is given by[7]

$$(a/a_E)^3 = \int_{-\infty}^{E} N(E)\, dE \Big/ \int_{-\infty}^{\infty} N(E)\, dE$$

According to the theory given here, σ_{min} can be observed in two ways:

(a) It is the lowest non-activated conductivity of a system in which $E_F - E_c$ can be varied.

(b) If E_F lies below E_c, the conductivity is

$$\sigma = \sigma_{min} \exp[(E_c - E_F)/kT] \tag{7}$$

in the high temperature range, though at low temperatures variable-range hopping will set it.

This theory, with its many approximations, is a theory of <u>non-interacting</u> electrons. If then it is desired to test it for instance through the behaviour of doped semiconductors, one should carry out experiments on conduction in an impurity band in strongly compensated semiconductors, for which the Hubbard U, the intra-centre interaction $\langle e^2/\kappa r_{12} \rangle$, does not play a major role. Experiments of this kind are those of the group at Lille[8, 9, 10] on n-type compensated InP and InSb, in which the transition to a non-metallic state is induced by a magnetic field; the field is thought to shrink the orbits, thus reducing $E_c - E_F$.

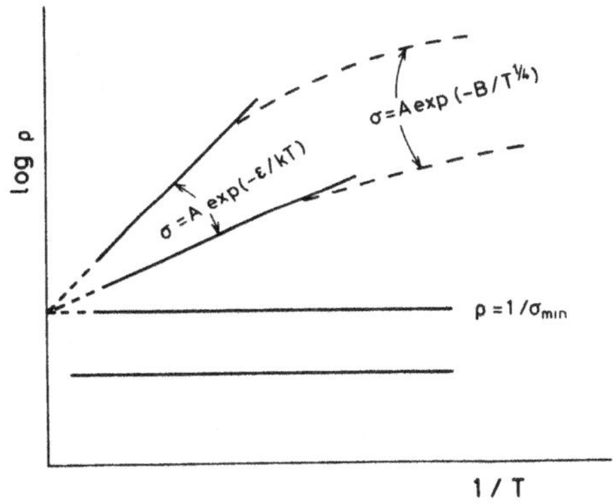

Figure 1: Plot of log (resistivity) against 1/T for varying values of $E_c - E_F$, changing sign at $\rho = 1/\sigma_{min}$. Such curves are observed for compensated semi-conductors with increasing magnetic field, or decreasing concentration.

Plots of log (resistivity) of the general type shown in figure 1 are found, the resistivity increasing with magnetic field up to about 0.9 T; the temperature range is between 1 and 10K. σ_{min} can be read off from the point at which the lines converge and it is found that the values obey (6), the best value of the constant being 0.027 for a range of $N_D - N_A$ from 6×10^{13} to 4×10^{16} cm^{-3}. Indeed there is much evidence from other systems that this is so over a range of a of about 10^2 (Pepper[11], Mott[12,13].) Of course this work does not show that a minimum metallic conductivity is a valid concept in the limit of low T; experiments at lower temperatures are urgently needed. But I think that, if values of σ (T = 0) much below σ_{min} do in fact exist, in the limit of low T, any theory must show why σ rises with T and flattens off at σ_{min}.

I will not comment on the scaling theories that predict that σ (T = 0) tends to zero as $(E - E_F)^{\nu}$, with in recent versions $\nu = 1$. With M. Kaveh[14] I have criticised the original version of Abrahams et al[15]. I want to remark, however, that if σ (T = 0) does vanish for extended states when $E = E_c$, then all the matrix elements

$\int \psi_1 (\partial/\partial x)\psi_2 \, d^3x$ between states at the Fermi energy, which occur squared in the Kubo-Greenwood formula, must vanish. I have argued[12] that this can only occur if the wave functions show long-range fluctuations, as illustrated in figure 2, and if the fluctuation length λ tends to infinity as $E - E_c \rightarrow 0$.

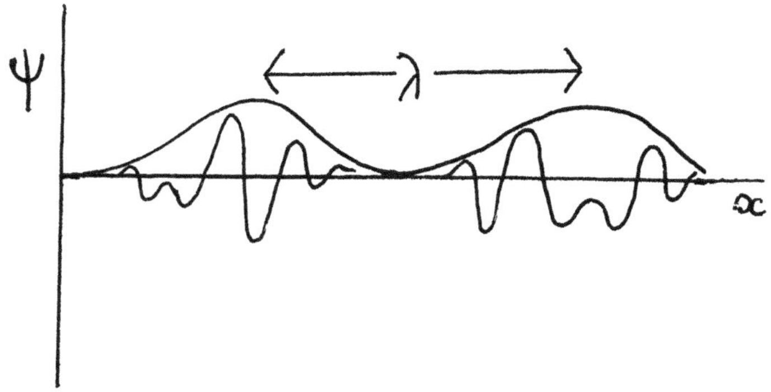

Figure 2: Showing schematically long-range fluctuations in a wave function.

That this might be so, by analogy with critical point behaviour, is suggested by Wegner (priv. comm.). If such fluctuations exist, it is easy to see that σ will rise to σ_{min} when the inelastic diffusion length L_i, due to electron-electron collisions or those with phonons, drops below λ . But according to a simple argument given by the present author[16, 12] and also by Stein and Krey[17] and by Sarker and Domany[18], these fluctuations depend on the value of the index s when we write the localization length $1/\alpha$ in the form

$$\alpha = \text{const}\,(E_c - E)^s,$$

Various values for s have been given, 0.6, $\frac{2}{3}$ (see reference 16) and from scaling theory the same value as ν for the conductivity, namely 1. So I see a contradiction here. Either the scaling theory or the above argument must be wrong.

I now turn to the experiments of Gordon Thomas and co-workers[19] on the metal-insulator transition in uncompensated silicon phosphorus. The quantity σ (T = 0), obtained by extrapolation to zero from measurements down to a few milli-Kelvin, shows the very sharp but not discontinuous drop shown in figure 3.

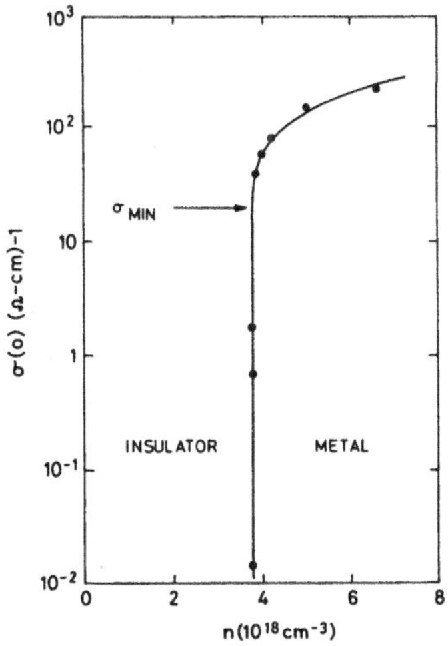

Figure 3: Conductivity in the limit as $T \to 0$ of uncompensated Si:P as a function of phosphorus concentration (Rosenbaum et al. 1980).

About these results I would make the following remarks.

1. If it were not for the random positions of the donors in space, giving very strong disorder, the transition, a Mott transition determined by the Hubbard U, would certainly be discontinuous,[20, 21.] Indeed, in fluid caesium and metal-ammonia solutions I have argued that the critical point is determined by this transition[20]. The degree of disorder is of course greater in Si:P, and I have until recently supposed that the disorder was great enough to wipe out the discontinuity (Mott[22]), and that the transition takes place in an impurity band. However, a striking result of Thomas[23] and coworkers is that, in the metallic regime, the electron's specific heat differs little from that calculated for free electrons. I therefore deduce that in the metallic regime electrons are in the conduction band, and that the transition is, here too, discontinuous, from a metallic state to one where electrons are localized.

2. The reason why σ drops so sharply arises from a correction to the conductivity first given by Berggren[24]

$$\sigma = \sigma_B \left[1 - \frac{3}{(k_F \ell)^2} \left\{ 1 - \frac{\ell}{L_i} \right\} \right]. \tag{8}$$

Here σ_B is the Boltzmann conductivity and ℓ the elastic mean free path. L_i once again is the inelastic diffusion length, or, in a magnetic field H, $(eh/Hc)^{\frac{1}{2}}$, whichever is the smaller. The equation is obtained from the Kubo-Greenwood equation by using wave functions corrected for diffusion, or in other ways. It is of course approximate, and should not be accurate near the zero of the term

$$1 - 3/(k_F \ell)^2, \tag{9}$$

where indeed one finds $\sigma \propto E - E_0$, where E_0 is the value at which (9) vanishes. This is in agreement with the scaling theory. But if one believes that σ goes continuously to zero, one has to ask again, can the matrix elements in the Kubo-Greenwood formula all vanish? Unless the long fluctuations of figure 2 exist, it is difficult to see how they could.

If then Si:P shows a discontinuous transition, σ as a function of n should appear as in figure 4; the smallest metallic conductivity will be that at which the discontinuous transition occurs, and has nothing to do with equation (6). I have discussed[7] how long-range fluctuations in potential existing in a random distribution may smear out the discontinuity, giving a curve as shown by the dotted line.

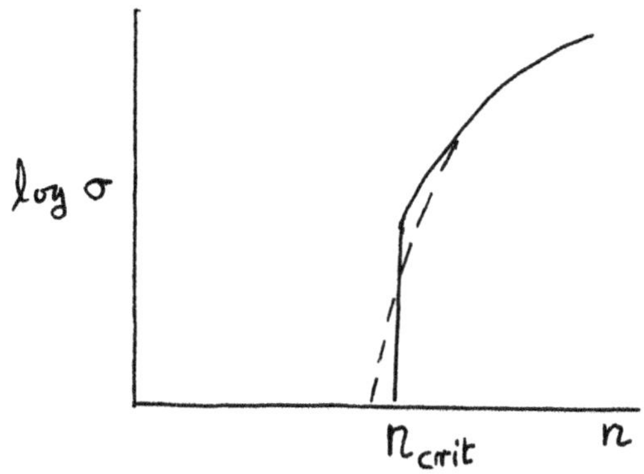

Figure 4: The expected behaviour of log σ in Si:P; if the transition is first order a discontinuity should be found even at a finite temperature. Long range fluctuations in potential may smear out the discontinuity as in the dotted line.

As regards the position of the mobility edge in a variety of materials, a density of states as in figure 5 is usually assumed. In amorphous semiconductors, many measurements (for a summary see reference 25) mainly of drift velocities give $\Delta E \sim 0.2$ eV for electrons in hydrogenated amorphous silicon, ~ 0.1 eV for holes in chalcogenides, and curiously no sign of a mobility edge for electrons in vitreous silica, and certainly not in liquid rare gases[6], for which we must assume $\Delta E < kT$. Spear and co-workers[26] in microcrystalline silicon heavily doped (> 19 cm^{-3}) with phosphorous, in which there are enough states in the gap to empty all the tail states, find $\Delta E = 0.07$eV. Davies[27] has attempted a calculation of ΔE for amorphous silicon, deducing it from the scattering cross section calculated for random dihedral angles and finds 0.25eV. A similar calculation for doped silicon should give ~ 0.1eV, as observed.

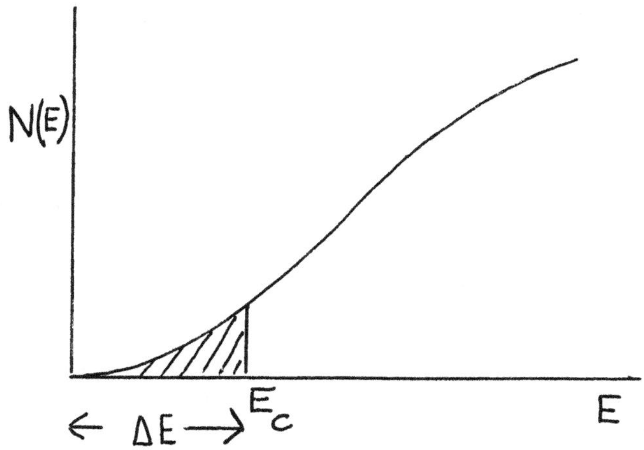

Figure 5: Density of states in a conduction band; E_c is the mobility edge.

A system in which the concentration of a degenerate gas of electrons can be varied, but in which the random field is due to negatively charged centres, is CeS (Cutler and Leavy[28], Mott and Davis[6] Chapter 4); there is here no chance of an impurity band and one would expect a straightforward Anderson transition as in (compensated) Si:P, except that here the electrons are in a conduction band, not an impurity band. This is what is observed, with σ_{min} in the range 10^{-2} to $10^{-3}\,\Omega^{-1}\,cm^{-1}$. There is no sign of the behaviour predicted by equation (8). Perhaps the temperature is too high (the measurements go down to 10K); but the electrons concentrations are high, greater than $10^{20}\,cm^{-3}$ at the transition. As the Fermi energy varies as $n^{2/3}$, one might suppose that this is $\sim$ 20 times greater than in Si:P at the transition, so 10K is equivalent to 0.5K in Si:P.

2. Application to liquids

In liquids all scattering processes are in principle inelastic (see Faber[29]). Therefore L_i in equation (8) is identical with ℓ, and the correction to the Boltzmann formula disappears. It seems therefore likely that these corrections to the conductivity in a conduction band are not applicable; a mobility edge should occur, and if $\Delta E > kT$ equation (7) should be valid.

If $\Delta E < kT$ we argue that the prefactor could be estimated as follows. The conductivity should be written

$$\sigma = ne^2\tau/m = \left\{ e^2 N(E) kT \ell / mv \right\} \quad e^{-W/kT}$$

where E = kT. For v we put in thermal velocities and suppose, following Ioffe and

Regel[30], that the mean free path cannot be less than the electron wave length $\hbar/mv$. On substituting we find for the pre-exponential factor a minimum value of

$$0.1\, e^2/\hbar\lambda_e \quad , \qquad \lambda_e = \hbar/mv. \tag{10}$$

At 600K, λ_e is 10^{-7}cm, so this quantity (10) is if anything rather greater than σ_{min}, if a_E is of order 10^{-7}cm.

For liquids the quantity $E_c - E_F$ is expected to be temperature-dependent. The usual way of obtaining the pre-exponential factor is through a plot of $\ln\sigma$ against the thermopower S; one expects

$$S = \frac{k}{e}\left\{\frac{E_F - E_c}{kT} + A\right\} \tag{11}$$

and thus

$$\log\sigma = \log\sigma_{min} - \frac{|S|(e/k) - A}{2.3}.$$

with A = 1 for conduction at a mobility edge. Results for fluid mercury (Schmutzler and Hensel[31]) and for fluid selenium (Hoshino et al[32]) give the correct slope as predicted but a small value of σ_{min} ($10 - 20\Omega^{-1}cm^{-1}$). So small a value is not understood. On the other hand for liquid Cs Freyland et al[33] find $300\Omega^{-1}cm^{-1}$, in agreement with (6).

Emin[34] first suggested that (11) is not correct if $E_F - E_c$ varies with T; according to him, if

$$-(E_F - E_c) = \varepsilon - \gamma T,$$

the thermopower should be given by

$$S = (k/e)\,[\varepsilon/kT + A]$$

Butcher and Friedman[35] came to the opposite view, but Emin[36] has recently defended his original viewpoint. If this is correct we should expect the plots of equation (11) to give a large pre-exponential factor $\sigma_{min}\exp(\gamma/k)$, which is as far as I know not observed.

As regards the metal-insulator transition, many years ago measurements of the Knight shift in In_2Te_3 and Ga_2Te_3 by Warren (refs. in 6) still provide, as far as I know, the most convincing evidence that $\sigma \propto g^2$ in the metallic range and that the transition occurs when $g = \frac{1}{3}$.

As regards metal-insulator transitions, of "Mott" type, we have already mentioned that for caesium and metal-ammonia the existence of a two-phase region indicates, in our view that the transition is in principle discontinuous. We think in metal-ammonia, for instance, that in the high concentration region a degenerate electron gas is present, and as the concentration decreases at temperatures above the critical point an impurity band forms and a transition of Anderson type occurs, the upper and lower Hubbard bands overlapping. At the transition $g \simeq \frac{1}{3}$ and a minimum metallic conductivity according to (6) is expected, and observed.

References

1. N.F. Mott, Phil. Mag. 26, 1015 (1972).
2. P.W. Anderson, Phys. Rev. 109, 1492 (1958).
3. J.T. Edwards and D.J. Thouless, J. Phys. C: Solid State Phys. 5, 807 (1972).
4. R. Kubo, Can. J. Phys. 34, 1274 (1950).
5. D.A. Greenwood, Proc. phys. Soc. 71, 585 (1958).
6. N.F. Mott and E.A. Davis, Electronic Processes in non-Crystalline Materials, Oxford (1979), Chap. 2.
7. N.F. Mott, Phil. Mag. B43, 941 (1981).
8. D. Ferré, H. Dubois and G. Biskupski, Phys. Stat. Solidi B70, 81 (1975).
9. G. Biskupski, H. Dubois and G. Ferre, Phil. Mag. B43, 183 (1981).
10. G. Biskupski, Thesis, Lille (1982).
11. M. Pepper, J. non-cryst. Solids 32, 161 (1978).
12. N.F. Mott, Phil. Mag., B44, 265 (1981).
13. N.F. Mott, Proc. R. Soc. A (in press) (1982).
14. N.F. Mott and M. Kaveh, J. Phys. C: Solid State Phys. 14, L659 (1981).
15. E. Abrahams, P.W. Anderson, D.C. Licciardello and T.V. Ramakrishnan, Phys. Rev. Lett., 42, 693 (1979).
16. N.F. Mott, Commun. Phys. 1, 203 (1976).
17. J. Stein and W. Krey, Z. Phys. B34, 287; ibid. 37, 13 (1980).
18. S. Sarker and E. Domany, J. Phys. C: Solid State Phys. 13, L273 (1980).
19. T.F. Rosenbaum, K. Andres, G.A. Thomas and R.H. Bhatt, Phys. Rev. Lett. 43, 1723 (1980).
20. N.F. Mott, Metal-Insulator Transitions, Taylor and Francis, London (1974).
21. W.F. Brinkman and T.M. Rice, Phys. Rev. B7, 1508 (1973).
22. N.F. Mott, Phil. Mag. B37, 377 (1978).
23. G.A. Thomas, Y. Ootuka, S. Koboyashi and W. Sasaki, Phil. Rev. B24, 4886 (1981).
24. K. Berggren, J. Phys. C: Solid State Phys. 15, L45 (1982).
25. N.F. Mott, J. Phys. C: Solid State Phys. 13, 5433 (1980).
26. G. Willeke, W.E. Spear, D.I. Jones and P.G. Le Comber, Phil. Mag. B (in press) (1982).
27. J.H. Davies, Phil. Mag. B41, 373 (1980).
28. M. Cutler and J.F. Leavy, Phys. Rev. A133, 1153 (1964).
29. T.E. Faber, Introduction to the Theory of Liquid Metals, Cambridge Univ. Press, (1972).
30. A.F. Ioffe and A.R. Regel, Prog. Semicond. 4, 237 (1960).
31. R.W. Schmutzler and F. Hensel, J. non-cryst. Solids, 8 - 10, 718 (1972).
32. H. Hoshino, R.W. Schmutzler and F. Hensel, Proc. 3rd Int. Conf. on Liquid Metals, Bristol (ed. R. Evans and D.A. Greenwood), p. 404 (1977). Inst. of Phys., London.
33. R.W. Schmutzler and F. Hensel, J. non-cryst. Solids 8 - 10, 718 (1972).
34. D. Emin, Proc. 7th Int. Conf. on Liquid and Amorphous Semiconductors (Edinburgh) ed. W.E. Spear, p. 249 (1977).
35. P.N. Butcher and L. Friedman, J. Phys. C: Solid State Phys. 10, 3803 (1977).
36. D. Emin, to be published.

Note added after the meeting:

Professor Hensel has pointed out that for mercury some recent investigations of conductivity and thermopower do indeed give the expected value for σ_{min} in the preexponential factor, namely $300\Omega^{-1}$ cm^{-1}, when the same T-dependent activation energy is taken for both quantities. The evidence is thus against Emin's conclusion for this case.

G. Schonherr, R.W. Schmutzler and F. Hensel, Phil. Mag. B49, 411 (1979).

I. IONIC LIQUIDS AND MOLTEN SALTS

IONIC LIQUIDS

F. Hensel

Institute of Physical Chemistry
University of Marburg
D-3550 Marburg, W.-Germany

I. Introduction

The study of ionic fluid systems has attracted interest for a long time for many technical and scientific reasons; and there is a very large literature on this subject extending back to the last century |1|. These systems are characterized by an appreciable concentration of ions, i.e. much of the cohesive energy of the liquids is Coulombic in nature and the diffusive motion of the ions can give rise to a noticable conductivity. There are several distinct groups of fluid systems which contain an essentially ionic component:

(a) Fully ionized salts like the alkali-metal halides have comparatively large values of the electronegativity difference between the constituent atoms. The energy band gaps of the solids at room temperature are larger than 3 eV and the electrical conductivity is mostly ionic under conditions of thermal excitation. The conductivity in the melt is entirely ionic; when electrolysed, fully ionized liquid salts conform to Faraday's law. The local conformation in the liquid is similar to that in the solid |2|.

(b) Slightly ionized molecular fluids $HgCl_2$, AlJ_3 or H_2O with intermediate electronegativity differences have energy band gaps sufficiently large that the electronic contribution to the electrical transport can be neglected. They have well-developed molecular structures at the melting point with a very small degree of dissociation. Application of pressure and high temperature shifts the self-ionization equilibrium. The poorly conducting slightly ionized liquids like e.g. $HgCl_2$ |3| and InJ_3 |4| or H_2O and NH_3 |5| become more ionic and consequently better conductors at high densities and high temperatures.

(c) A very interesting group is formed by those liquids that contain an ionic component even in the absence of acceptors, i.e. with definitely no difference in the electronegativity. A well known example is liquid iodine which forms at its melting point a predominantly molecular liquid with a very small degree of self-ionization. The ionization equilibrium ($2J_2 \rightleftarrows J^+ + J_3^-$) |6|, |7| is again shifted in favour of the right side by an increase in density |4|, |7|.

(d) In recent years a considerable amount of work has centered on the group of liquid conductors which are at the borderline between metals, electronic semiconductors and ionics. This includes e.g. the liquid mixtures Mg-Bi, Li-Pb, Tl-Te, Li-Bi and Cs-Au. The mixtures have quite a low conductivity in the neighbourhood of compositions satisfying simple chemical valence requirements |8|, |9|. Whenever available, the thermochemical data, especially the excess Gibbs free energies, indicate that chemical compounds form (cp. e.g. CsAu |10|, Li_4Pb |11|, Mg_3Bi_2 |12| and Tl_2Te |8|). Support for the character of these alloys being at least partly ionic stems often from the liquid structure which has been investigated by X-ray and neutron diffraction techniques. The structure of these melts is characterized by short-range ordering similar to that in fully ionized liquid salts (cp. e.g. CsAu |13|, Li_4Pb |14|, Tl_2Te |15|). These liquid compounds exhibit features of both electronic and ionic conduction depending on the magnitude of the gap. The conductivity of molten CsAu e.g. is predominantly ionic; when electrolysed, it conforms to Faraday's law.

(e) Another group of fluids in which Coulombic interaction between charged particles plays an important role is formed by plasmas of relatively low temperatures (10^3K < T < 10^4K) but considerably high pressures (10^2bar < p < 10^4bar), i.e. at liquid-like densities. These systems are clearly of interest because of the possible existence of plasma phase transitions which are connected with ionization phenomena and many particle interactions |16|, |17|. Experimental research on such systems is rather new |18|. It will be extensively discussed in this paper.

A more extended review of work on systems (a) to (d) is unnecessary because reference may be made for further details to a large number of surveys that exist (see e.g. for (a)-(d) |9|, for (a) |19|, |20|, |21|, for (b) |5|, |4|, |22|, for (d) |8|, |23|, |24|, |25|. A rather new development in the field of fully and partly ionized salts is experimental |26|, |27| work on the nature of the vapour-liquid critical point phase transition. Especially, the supercritical phase shows remarkable electrical behaviour. A continuous transformation from a highly conducting ionic fluid to a non-conducting molecular vapour can be obtained |27|.

In the following I will select for attention those electronically conducting fluid systems which are at the borderline between fluid metals, fluid semiconductors or fluid ionics ((d) and (e)). It has been generally recognized that for these fluids the difficulty in precisely defining the boundaries between the different types of conduction is intimately related to the existence of a metal-to-nonmetal transition which is observed on changing a thermodynamic variable such as composition density or temperature. Therefore, in what follows I will restrict the presentation to a few results with special emphasis on the following questions: What is the role of charge transfer for the characteristics and mechanisms of the metal-to-nonmetal transitions? What is the correlation between the metal-nonmetal transition and the phase behaviour of these fluids? Where in the phase diagram does the transition occur? And is a breakdown of microscopic homogeneity connected with this transition?

I believe that the best way to approach the problem is to concentrate the discussion on two extreme examples: Liquid Cs-Au with a relatively high difference in the electronegativities of the constituents and a liquid compound CsAu known to be ionic in character, and fluid mercury, for which the electronegativity difference is definitely zero. Fluid mercury is of special interest, because of very recent experimental results |18| which we regard as compelling evidence that the ground state of the expanded fluid just below the metal-nonmetal transition density contains an ionic component even in the absence of electron-acceptor-constituents.

II. The Metal-Nonmetal Transition in Cs-Au

A comprehensive review of the existing work on CsAu is unnecessary because of the surveys that exist |9|, |25|. Recent new studies of the electron localization mechanism in course of the metal-nonmetal transition and of the thermodynamic stability of the liquid compound will be discussed by Warren and Weil in this Proceedings. Therefore, it is sufficient to summarize in the following a few noteworthy observations.

The occurrence of a concentration induced gradual metal-nonmetal transition in liquid Cs-Au is clearly established by the well known behaviour of the electrical conductivity σ as shown in fig.1. σ changes by almost 4 orders of magnitude and has a pronounced minimum of 3 $ohm^{-1}cm^{-1}$ at the stoichiometric composition CsAu. The transition to nonmetallic behaviour is accompanied by a change in the temperature dependence from

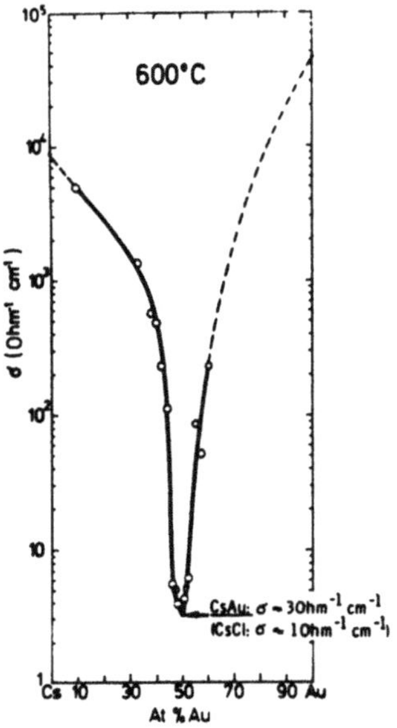

Fig.1: Electrical conductivity σ of liquid Cs-Au mixtures as a function of the gold content at 600°C.

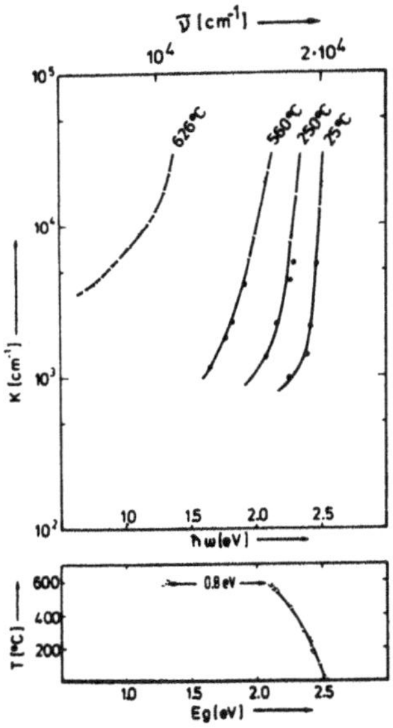

Fig.2: Top: Optical absorption of a 100 µm-thick sample of solid or liquid CsAu as a function of the photon energy. Bottom: The temperature dependence of the position of the absorption edge. The arrow indicates the abrupt change of the position of the edge at the melting point (590°C).

negative to positive at 44 at% Au corresponding to a σ of about 100 $ohm^{-1}cm^{-1}$. The corresponding apparent activation energy approaches a maximum of about 0.2 eV. This value is a factor of about 7 smaller than the optical energy gap in liquid CsAu of 1.3 eV taken from fig.2. Fig.2 |28| shows the spectral dependence of the absorption coefficient K(ω) for solid CsAu (from 25°C to 560°C; CsAu melts at 590°C) and for liquid CsAu (626°C) as a function of the photon energy. The absorption edges of the solid phase exhibit the expected red shift which is usually observed for ionic insulators with increasing temperature. On melting, there is an abrupt change in the position of the absorption edge of about 0.7 to 0.8 eV. The lower part of fig.2 shows the temperature dependence of the position of the optical gap which corresponds roughly to the gap between valence- and conductionband. It is a plot of the photon energy corresponding to $K = 10^4\ cm^{-1}$ as a function of the temperature. There is a striking quantitative similarity between the behaviour of CsAu and that of molten salts. Direct experimental evidence that liquid CsAu is entirely ionic like fully ionized salts stems from an electromigration experiment |29|. When electrolyzed, liquid CsAu conforms to Faraday's law.

The concentration dependence of the conductivity σ in the liquid Cs-Au system (fig.1) establishes the occurrence of a transition from metallic to nonmetallic behaviour. At σ-minimum for the stoichiometric composition CsAu the electrical transport is determined by the diffusive motion of Cs^+ and Au^-. Therefore, it seems reasonable to assume that bound states associated with the formation of Cs^+Au^- ionic assemblies remove electrons from the conduction band of cesium if the concentration is changed between pure Cs and CsAu; i.e. the liquid alloys over the whole concentration range - including the metallic range - are assumed to be binary mixtures of Cs^+ and Au^- and the excess metal Cs. Direct evidence for the validity of this assumption comes from recent measurements of the total structure factor S(q) by Martin et al. |13| . Fig.3 shows S(q) together with the total pair correlation function g(r) for different liquid Cs-Au mixtures which were obtained from neutron diffraction data and which are reproduced from reference |13| . As pointed out by Martin et al. |13| , S(q) and g(r) for the equiatomic CsAu-melt cannot be described with the hard sphere model for the statistical distribution of atoms of both kinds |30| . They can, however, reasonably be modelled by charged hard spheres |31| ; both theory and experiment exhibit a pronounced shoulder on the low-angle side of the main peak of S(q). This is strong evidence for a molten salt-like charge ordering in CsAu. In view of the hypothesis that an Au-ion is formed even in very dilute

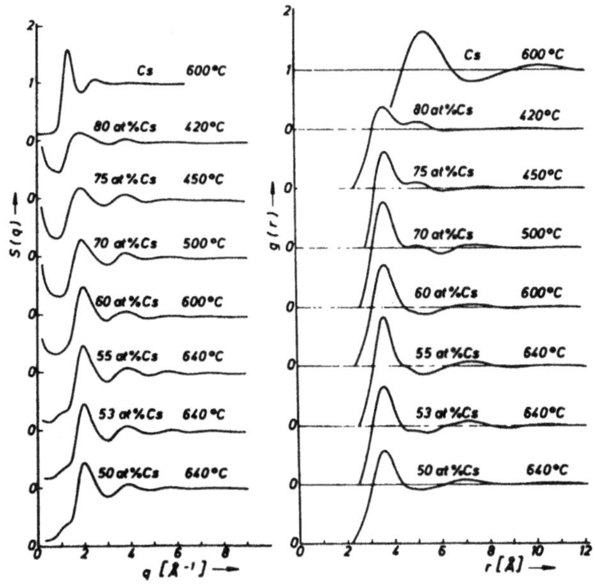

Fig.3: Total structure factors and total pair correlation functions of Cs-Au melts |13|.

solutions of Au in Cs it is especially interesting that for an Au-concentration of only 20 atom-% a first maximum is observed at 3.55 Å. At this concentration the conductivity is still higher than 10^3 $ohm^{-1}cm^{-1}$ and in the metallic range. This is a clear indication that Cs^+ and Au^- are formed even in the diluted solution. Martin et al. |13| showed that the concentration dependence of the heights and the shapes of the first two peaks in g(r) and, in particular, the small angle scattering for $q < 1$ $Å^{-1}$ are consistent with the existence of a segregation tendency in course of the metal-to-nonmetal transition, i.e. with the existence in the average of time of regions containing salt-like Cs^+Au^- and those containing metallic Cs. The assumption of such a segregation effect is certainly consistent with recent measurements of the concentration dependence of the Gibbs excess free energy of mixing of liquid Cs-Au alloys by A.Kempf |10| (cp. also the contribution of K.Weil) from which in the usual way |32| the mean-square fluctuations in the concentration $N\langle\Delta x^2\rangle$ (N = the number and x = mole fraction) can be obtained. The results in fig.4 again indicate a high stability of the liquid compound CsAu and large concentration fluctuations in course of the metal-to-nonmetal transition in Cs-rich mixtures. It thus appears to us that microscopic "segregation" is necessarily connected with the metal-to-nonmetal transition in fluids. This effect must be intimately connected with the long range Coulomb forces that play an important role in systems containing electrons and ions. The next section presents experimental strong evidence that the macroscopically observed metal-to-nonmetal transition in fluid Hg is preceded by a "microphase" separation.

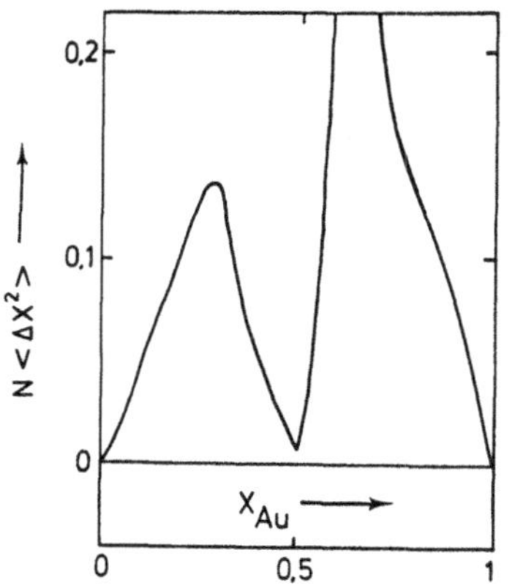

Fig.4: Mean-square fluctuations in the concentration of liquid Cs-Au mixtures. The temperature is 604°C.

III. Plasma Transition and Metal-Nonmetal Transition in Fluid Mercury

It is well known, that in recent years, both the demands of new technologies and continuing interest in the development of adequate theories have led to a great increase in research into the properties of fluid metals over a wide range of temperatures up to supercritical temperatures and over a wide range of density from the dense liquid to the dilute vapour. Because liquid mercury has a relatively low critical temperature (T_c = 1750 K, p_c = 1670 bar, d_c = 5.77 g/cm^3), extensive experimental results including dc conductivity, thermoelectric power, Hall effect, Knight shift, optical reflectivity and -absorption, and thermodynamic data are available at sub- and supercritical conditions. Despite this large amount of work there still remains a fundamental controversy about the nature of the vapour-liquid critical point phase transition of mercury. The theoretical treatment of this phase transition is especially difficult in metals due to the absence of an adequate interatomic potential function for the entire liquid-vapour range. Regardless of the way in which the forces in a metal are described the description must change with density. It is known that a change of state from point L to point G in fig.5 across the vapour pressure curve, at relatively low temperatures far way from the critical point, c.P., is accompanied by a discontinuous transition from metallic to nonmetallic behaviour. This transition, however, implies that the type of the interatomic cohesion changes with temperature and pressure along the dashed curve round c.P. in fig.5.

At low temperatures, near the triple point, the interaction in the liquid is metallic characterized by the large electrical and thermal conductivity of the free electron gas, whereas in the low density vapour the majority of electrons are attached to their parent atoms and molecules which interact via van der Waals' forces. However, this assumption is certainly no longer valid at temperatures close to the critical

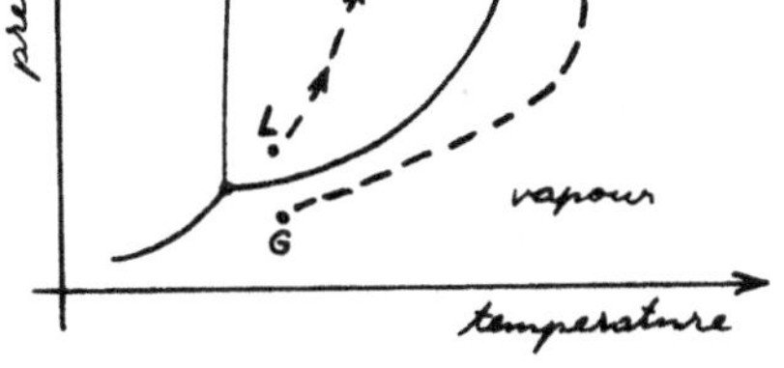

Fig.5: The liquid-vapour equilibrium and the critical point, p-T-diagram.

temperature. Under such conditions thermal ionization effects must become significant in the dense vapour phase. Consequently, in addition to the van der Waals interaction between neutral particles, the charge-neutral interaction and the Coulomb interactions between charges can play an important role. Since all these interactions lead to volume and temperature dependent contributions to the free energy of the fluid, it is certainly tempting to speculate about the existence of different plasma phase transitions. The possibility exists that under strongly non-ideal conditions, when the interaction energies of the charged particles among themselves or the interaction energies between the charge and the neutrals become comparable with the energy of their thermal motion the plasma splits into two phases, weakly and highly ionized, each one in ionization equilibrium. On the other hand, if the density of the fluid is increased still further along the dotted line in fig.5 a transition to a metallic state must occur in some regions of the p-T-diagram.

The large amount of work devoted over the last 15 years to the experimental and theoretical investigation of the metal-to-nonmetal transition in fluid mercury has been fully reviewed |25|, |33|, |34| and need not detain us here. It is sufficient to summarize a few noteworthy observations.

The best way to approach the problem is to measure simultaneously the electrical conductivity σ and the density as a function of pressure and temperature T. Fig.6 gives a selection of such data for mercury. Apart from the liquid-vapour phase transition (qualitatively indicated by

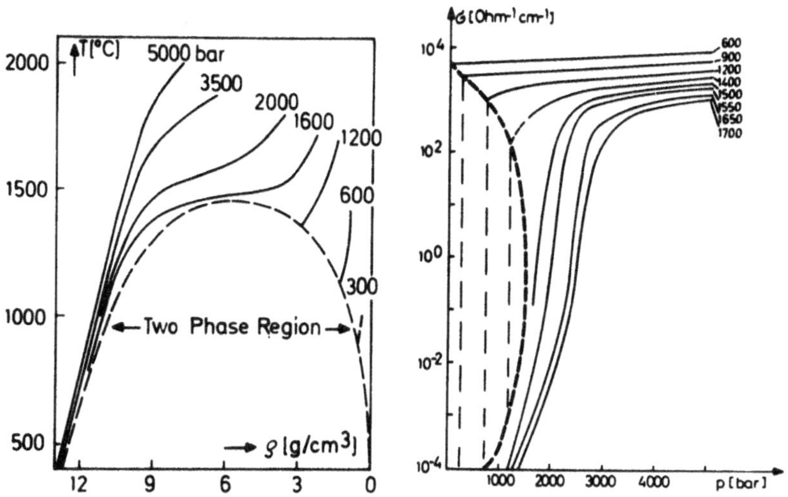

Fig.6: Equation of state and electrical conductivity of fluid mercury.

dotted lines) no discontinuous changes are indicated in σ or d. However, it must be pointed out that the possibility of narrow additional discontinuities cannot be completely ruled out because of the inaccuracy of the existing data (fig.6). At the critical point the conductivity of mercury is with less than 10^{-1} $ohm^{-1}cm^{-1}$ relatively small and certainly nonmetallic. All the experimental data (including optical data, electrical conductivity, Hall effect, thermopower, and Knight shift) show a gradual diminution of metallic properties in liquid mercury if the temperature is increased or the density is decreased along the coexistence line. Somewhere around T = 1660 K which corresponds to a liquid density under saturation condition of about 9 g/cm^3 Hg is nonmetallic. This behaviour is nicely illustrated by the optical conductivity of liquid Hg along the coexistence line as shown in fig.7. In the free electron metallic range for high densities the low frequency absorption is Drude-like. A gradual change from metallic to nonmetallic behaviour occurs in the density range between 11 and 9 g/cm^3. For still smaller densities a real gap seems to open. The metal-nonmetal transition in Hg occurs wholly in the liquid phase. For densities smaller than 9 g/cm^3 (i.e. temperatures larger than 1660 K the liquid-vapour phase change is from a nonmetallic liquid to a less conducting gas.

Let us turn next to the plasma state of the dense Hg-vapour close to the critical point. Our knowledge of this region lags far behind those of the high density liquid and the low density vapour. This is not surprising because of the extreme experimental problems associated with measurements in that region. However, the author's view is that the know-

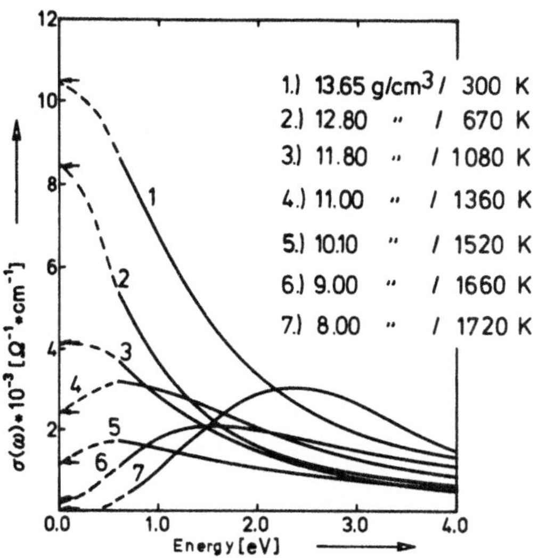

Fig.7: Conductivity of liquid mercury as a function of the photon energy at different densities (temperatures)

$$\sigma(\omega) \propto \int \frac{|D|^2 N(E)\, N(E+\hbar\omega)}{\hbar\omega}\, dE$$

ledge of the behaviour of this dense vapour phase of Hg is particularly important for the understanding of the nature of the vapour-liquid phase transition and the mechanism of the metal-to-nonmetal transition in fluid Hg. Therefore, in the following we will discuss very recent optical results |18| which indicate that the vapour liquid phase transition in Hg near its critical point is preceded by a "plasma transition".

Optical reflectivity and absorption measurements have been used to determine the absorption coefficient and the real part of the dielectric constant ε_1 in the infrared spectral range $0.5 < \hbar\omega < 1$ eV. Selected experimental results for ε_1 at the constant infrared energy $\hbar\omega = 0.6$ eV as a function of pressure are shown in fig.8 for the subcritical and supercritical temperature T = 1710 K and T = 1800 K, resepctively. The arrows p_V (1710 K) and p_c denote the vapour pressure at 1710 K and the critical pressure of Hg, respectively. The characteristic feature of these curves is the sharp enhancement of ε_1 by a factor of roughly 4 at pressures of about 1400 bars. This dielectric anomaly occurs clearly in the superheated unsaturated vapour phase far away from the vapour-liquid phase line. It is extremely sharp at the subcritical temperature, and it is smeared out at the supercritical temperature. The increase of ε_1 is followed by a smooth decrease in the direction to negative values.

When ε_1 is plotted as a function of density d, as e.g. in fig.9 for the subcritical temperature 1710 K, three distinct density ranges are apparent: for $d < 3$ g/cm^3 the density-dependence of ε_1 can be to a first approximation described by the Clausius-Mosotti relation for $\hbar\omega = 0$ using the polarizability of the isolated Hg atom (dashed line C.M.); around a

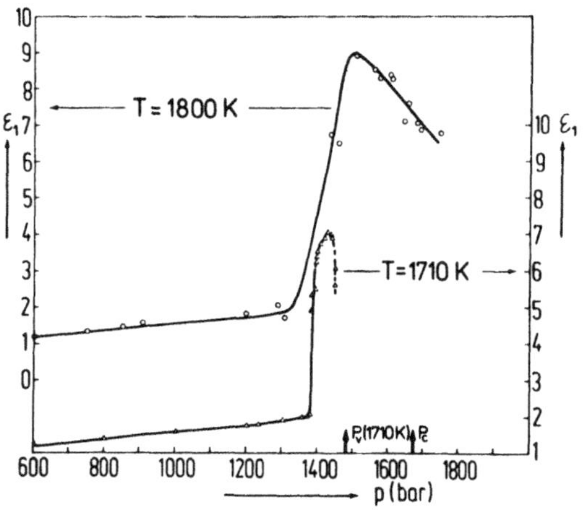

Fig.8: Real part of the dielectric constant ε_1 of Hg-vapour at the constant infrared energy $\hbar\omega = 0.6$ eV as a function of pressure p for the sub- and supercritical temperature T = 1710 K and 1800 K, respectively. The arrows indicate the vapour pressure p_V at 1710 K and the critical pressure p_c.

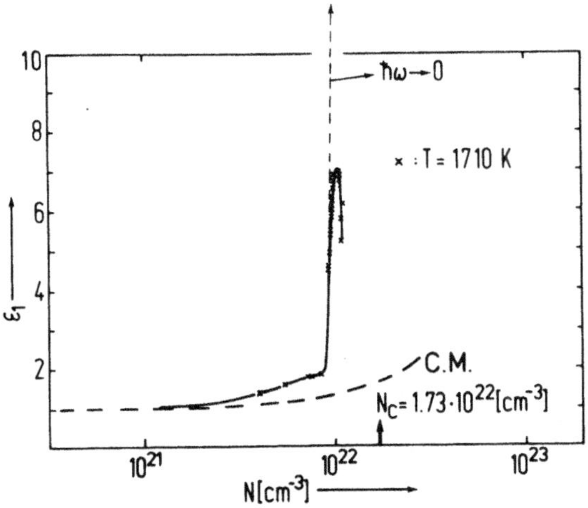

Fig.9: Real part of the infrared dielectric constant ε_1 of mercury vapour at 1710 K as a function of the number density N ($d = 1\ g/cm^3 = N \triangleq 3\cdot10^{21}\ cm^{-3}$). The dashed line (C.M.) is the Clausius-Mosotti relation. N_c is the critical density.

density of about 3 g/cm^3 a striking upward deviation from Clausius-Mosotti behaviour is observed in an extremely narrow density range; finally there is a smooth decrease of ε_1 at larger densities. The onset of the dielectric anomaly is accompanied by an abrupt change of the character of the optical absorption and emissivity spectrum. In the low density range, a line-spectrum is observed with the main absorption lines at 4.89 eV and 6.7 eV corresponding to transitions between the 6s ground state and the 6p triplet and singlet level of the Hg-atom. With increasing vapour density the sharp lines begin to broaden and to be superimposed on a background absorption resulting in a relatively steep absorption edge which moves rapidly to lower energies with increasing density. For $d < 3\ g/cm^3$ the edge has two distinct regimes: a nearly exponential increase of the absorption with photon energy becomes quadratic at higher photon energies. These two regimes have been recently successfully interpreted in terms of absorption by excitonic states of large randomly distributed nonmetallic Hg clusters |35|. For $d > 3\ g/cm^{-3}$ abruptly a third absorption regime appears. Strongly density- and temperature dependent absorption tails extend in the infrared spectral range. At the same time the dc conductivity remains very small (10^{-3} $ohm^{-1}cm^{-1}$). From a careful analysis of reflectivity, absorption and dc conductivity the spectral dependence of the optical conductivity of Hg-vapour at $d > 3\ g/cm^3$ (shown schematically in fig.10) can be derived. It is easily seen from an analysis of $\sigma(\omega)$, applying the Kramers-Kronig dispersion relation for ε_1 (eq.1), that the density induced increase of ε_1, which is smeared out over a certain density range at the finite frequency $\hbar\omega = 0.6$ eV, becomes extremely sharp and more enhanced in the very low frequency limit $\hbar\omega \rightarrow 0$ as indicated by the dotted line in figure 9.

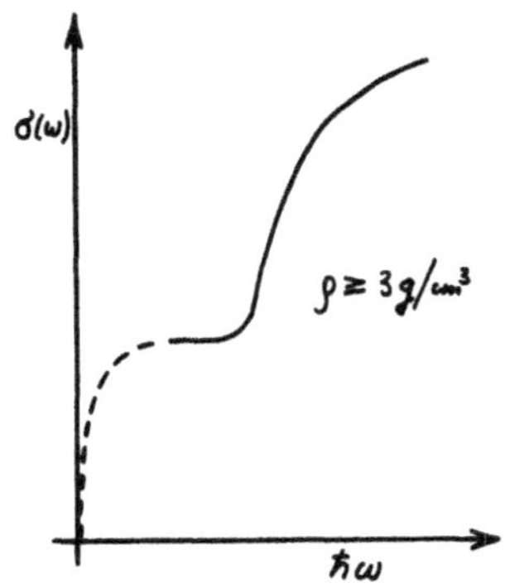

Fig.10: Schematic illustration of the shape of the optical conductivity-curve of Hg-vapour for densities larger than 3 g/cm^3.

$$\varepsilon_1(\omega) = 1 + 8\ P \int_0^\infty \frac{\sigma(\omega')d\omega'}{\omega'^2 - \omega^2} \qquad (1)$$

(where P denotes the principal part).

We believe that the abrupt change of the optical properties of Hg at $T = 1710$ K for $N > 9\cdot 10^{21}$ cm^{-3} and, in particular the sharp rise of ε_1 indicate that the interaction of thermally generated electrons and ions with neutral atoms stabilize regions of higher than average density, especially heterophase fluctuations in the compressed vapour phase.

There are a few theoretical papers which support the hypothesis that under certain conditions of the vapour charges tend to stabilize dense liquid droplets which approach metallic conditions while the macroscopic system is still in the nonmetallic region. The first reference that we have seen is given by Lifshitz and Gredescul |36|, who showed by an analysis of the free energy of a system consisting of an electron and a medium of classical particles that the appearance of an indirect particle interaction via the electron can cause a transition from a normal gas distribution to a dense cluster in a narrow interval of T or N, respectively.

A second approach to the problem is a computer experiment by Lagarkov and Sarychev |37| which examined electrons, obeying Newton's laws with a semiclassical density of states, interacting with a set of neutral Hg-atoms. The position of the scatterers was kept random, with the additional constraint of a hard-core interaction between the Hg-atoms. The electron-scatterer interaction was parametrized to simulate the electron-neutral mercury atom interaction. The results again were interpreted as electron localization, at low energies, on clusters of mercury atoms. An analysis of the spatial correlation function for the electron-neutral Hg interaction showed that the density of the cluster can approach values characteristic of the liquid, i.e. larger than the critical density.

Very recently, Hernandez |38| has performed a quantum mechanical, variational calculation to probe the requisite conditions for electron stabilization of high density fluctuations in fluid Hg. The onset of stabilization has been found to occur when electrons are self-consistently bound to small volumes with an excess density at the cluster center of at least 50% more than average. The calculated density-temperature threshold for the onset of the electron-cluster stability is in qualitative agreement with the experimentally observed dielectric anomaly. A very recent approach by Hefner et al. |39| has been undertaken to probe whether the attractive forces between the mercury atoms within the cluster can lead to an abrupt increase in the density, i.e. to an "analogue of the vapour-liquid phase transition" within the cluster. Instead of attempting to make an exact calculation of the attractive cohesive energy in the cluster a mean field approximation was applied which assumes that the fluid consists of hard-sphere particles moving in an uniform potential field. The underlying idea of this treatment is that in the absence of charges the equation of state and the vapour-liquid phase transition of fluid Hg can be qualitatively modelled by the van der Waals theory. The result of this calculation is, that the conditions for heterophase fluctuations is not met at low densities and high temperatures. However, for densities of about 3 g/cm^3 (at which the onset of the dielectric anomaly is observed) the model predicts the formation of stable liquid droplets with a central density of about 9 g/cm^3, the value required for the nonmetal to metal transition in liquid mercury |33|.

We remark finally on the location of the plasma transition in the phase diagram of fluid Hg in the pressure-temperature- and density-temperature plane (fig.11). On the basis of the foregoing discussion we assume that across the dotted line a "transition" occurs from a normal slightly ionized gas phase to an equilibrium disperse system containing charged metallic droplets. Of special interest is the point marked T_x which subdivided the liquid-vapour coexistence line into two parts. For $T < T_x$ a transition from a normal rarefied vapour to a microscopically homogeneous metallic liquid is observed. As described above (cp. e.g. the behaviour of $\sigma(\omega)$ in fig.7) with increasing T a gradual diminution of metallic properties is observed in liquid Hg along the coexistence line. At $T = T_x$ all properties of the liquid change over smoothly to distinctly nonmetallic behaviour. The vapour-liquid phase transition for $T > T_x$ thus appears to us as a condensation of a vapour containing disjoint stable charged metallic clusters interacting via screened Coulomb for-

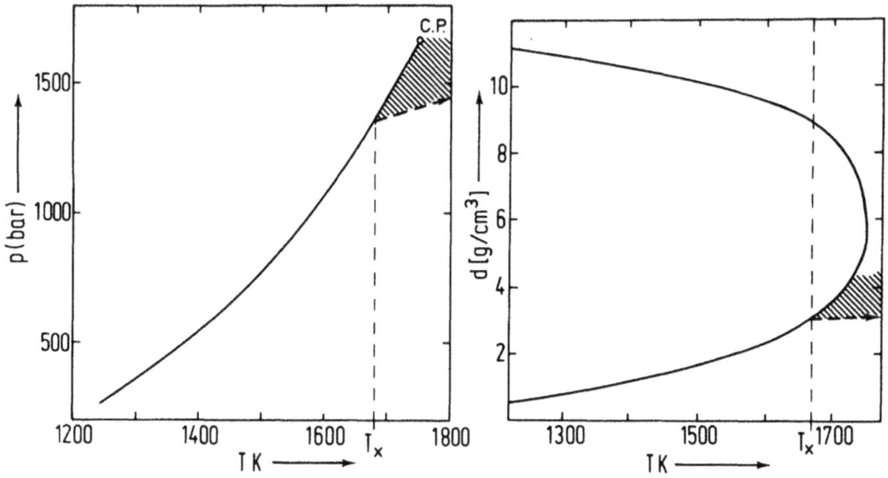

Fig.11: The phase diagram of liquid mercury in the p-T- and d-T plane. The dotted lines indicate the location of the dielectric anomalies at subcritical temperatures.

ces to a liquid which retains macroscopically a nonmetallic character. The change in the character of the vapour liquid phase transition at $T = T_x$ leads to a distortion of the liquid-gas phase diagram. An analysis of the existing liquid and vapour densities along the coexistence curve by Schmutzler shows a clear change of the shape of the coexistence curve at about T_x.

References

1 Faraday,M., Experimental Researches in Electricity I, (London: Richard and John Taylor, 1849)
2 Edwards,F.G., Enderby,J.E., Howe,R.A., and Page,D.J., J.Phys.C 8, 3483 (1975)
3 Bardoll,B. and Tödheide,K., Ber.Bunsenges.Phys.Chem. 79, 490 (1975)
4 Cleaver,B., Spencer,P.N., and Quddus,M.A., J.Chem.Soc.Faraday Trans. I 74, 686 (1978)
5 Franck,E.U., Ber.Bunsenges.Phys.Chem. 76, 341 (1972)
6 Beacroft,D.J. and Nachtrieb,N.H., J.Phys.Chem. 71, 316, 4400 (1967)
7 Böttger,D., Diplom-Thesis, Universität Marburg (1978)
8 Cutler,M., Liquid Semiconductors (New York: Academic Press, 1977)
9 Hensel, F., Adv.Phys. 28, 555 (1979)
10 Kempf,A., Thesis, Universität Marburg (1981)
11 Saboungi,M.-L., Marr,J., and Blander,M., J.Chem.Phys. 68, 1375 (1978)
12 Egan,J.J., Acta metall. 7, 560 (1959)
13 Martin,N., Freyland,W., Lamparter,P., and Steeb,S., Phys.Chem.Liquids 10, 61, 77 (1980)
14 Ruppersberg,H. and Egger,J., J.Chem.Phys. 63, 4095 (1975)
15 Gay,M., Enderby,J.E., and Copestake,A.P., to be published 1982

16 Ebeling,W., Kraeft,W.D., and Kremp,D., Theory of Bound States in Plasmas and Solids (Akademic-Verlag, Berlin, 1976)

17 Norman,G.E. and Starostin,A.N., Teplofiz. Vys.Temp. 8, 413 (1970)

18 Hefner,W. and Hensel,F., Phys.Rev.Letters 48, 1026 (1982)

19 Klemm,A., Molten Salt Chemistry, ed. by M.Blander (New York: Wiley, 1964)

20 Angell,C.A., Rev.Phys.Chem. 22, 429 (1971)

21 Copeland,J.L., Transport Properties of Ionic Liquids (New York: Gordon and Breack, 1974)

22 Tödheide,K., Proceedings of the International Symposium on Molten Salts (Princeton: The Electrochemical Society, 1976)

23 Enderby,J.E., Amorphous and Liquid Semiconductors, ed. J.Tauc (New York: Plenum Press, 1974, p.361)

24 Warren,W.W., Advances in Molten Salt Chemistry, Vol.4 (1981)

25 Hensel,F., Angew.Chem.Intern.Edit. 19, 593 (1980)

26 Buback,M. and Franck,E.U., Ber.Bunsenges.Phys.Chem. 76, 350 (1972), 77, 1074 (1973)

27 Treiber,G. and Tödheide,K., Ber.Bunsenges.Phys.Chem. 77, 540, 1079 (1973)

28 Münster,P. and Freyland,W., Phil.Mag. B39, 93 (1979)

29 Krüger,K.D. and Schmutzler,R.W., Ber.Bunsenges.Phys.Chem. 80, 816 (1976)

30 Ashcroft,N.W. and Langreth,D.C., Phys.Rev. 156, 685 (1967)

31 Evans,R.E. and Telo da Gama,M.M., Phil.Mag. 41, 351 (1980)

32 Bathia,A.B. and Thornton,D.E., Phys.Rev. A2, 3004 (1970)

33 Mott,N.F., Metal-Insulator-Transitions (London: Taylor and Francis, 1974)

34 Cusack,N.E., in Metal Non-Metal Transitions in Disordered Systems, ed. L.R.Friedman and D.P.Tunstall (Edinburgh, 1978)

35 Bhatt,R.N. and Rice,T.M., Phys.Rev. B20, 466 (1979)

36 Lifshitz,I.M. and Gredescul,S.A., Zh.Eksp.Teor.Fiz. 57, 2209 (1969)

37 Lagarkov,A.N. and Sarychev,A.K., Teplofiz.Vys.Temp. 16, 903 (1978)

38 Hernandez,J.P., Phys.Rev.Letters, to be published 1982

39 Hefner,W., Sonneborn-Schmick,B., and Hensel,F., Ber.Bunsenges.Phys. Chem., to be published.

THE METAL-NONMETAL TRANSITION IN IONIC LIQUIDS

W. W. Warren, Jr.

Bell Laboratories

Murray Hill, New Jersey 07974

ABSTRACT

The metal-nonmetal transition in ionic liquids is reviewed with special emphasis on alkali metal-alkali halide solutions and the ionic alloy Cs-Au. Magnetic measurements, especially nuclear magnetic resonance, are discussed in relation to the nature of localized electronic states at low concentrations of excess metal and electron dynamics in the metal-nonmetal transition region. It is suggested that the stability of non-magnetic species governs the occurrence of a continuous metal-nonmetal transition or, alternatively, liquid-liquid phase separation.

I. INTRODUCTION

Solutions of liquid metals and ionic liquids exhibit fascinating electronic and thermodynamic properties (1-4). Some of these systems offer the opportunity to study the physics of low densities of conduction electrons influenced by strong ionic potentials and also continuous transitions from the metallic to the nonmetallic state as the composition is varied. However the possibility of intermediate compound formation or liquid-liquid phase separation leads to a rich variety of behavior in different solutions (1). We can identify three main groups for which typical temperature-composition phase diagrams are illustrated in Fig. 1.

(i) *Alkali metal-alkali halide solutions.* Most solutions of liquid alkali metals with their molten halides exhibit a range of liquid-liquid immiscibility below a critical temperature T_c. However, as shown in Figs. 1a-1c, there is a systematic trend toward lower values of T_c (relative to the melting temperature) on passing from Na solutions, Na-NaX, to the heavier metals. The Cs-CsX solutions, in fact, display no macroscopic phase separation although thermodynamic investigations indicate concentration fluctuations characteristic of values of T_c below the liquidus (5,6).

(ii) *Divalent and polyvalent metal-halide solutions.* These systems often form stable 'subhalides' based on metal ions of lower oxidation state, i.e. Hg_2Cl_2, InI, etc. In such cases there is little mutual solubility between the metal and the 'subhalide' and the T_c values are very high (Fig. 1d). In a few cases such as Ba-BaX_2 and Bi-BiX_3, the 'subhalide' is unstable and the region of phase separation is correspondingly reduced (Fig. 1e).

(iii) *Metal-metal solutions (alloys).* A number of liquid alloys involving a metal and a chalcogen (Fig. 1f) or two liquid metals such as Cu-Au (Fig. 1g) exhibit electronic properties characteristic of semiconductors or molten salts near compositions of a well-defined stoichiometry (37). The compositions usually correspond to satisfaction of simple valence requirements (Ga_2Te_3, Tl_2Se, CuAu, Li_4Pb, etc.). Most metal-chalcogen alloys exhibit a narrow range of phase separation between the metal and the nonmetal; some show liquid immiscibility on the chalcogen side as well. The Cs-Au system is noteworthy for the similarity of its phase diagram in the range Cs-CsAu with that of Cs-CsX (Figs. 1c and 1g). This system is discussed by Hensel and by Weil elsewhere in these Proceedings.

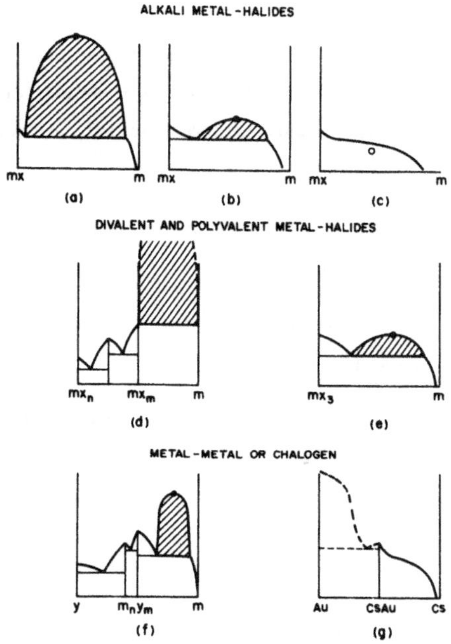

Figure 1. Schematic temperature-composition phase diagrams. Alkali metal (m) - alkali metal halides (mx): (a) m=Na, (b) m=K or Rb, (c) m=Cs. Divalent and polyvalent metal (m) - halides mx_n: (d) m=Cd, Hg, Ga, In, (e) m=Ca, Ba, Bi. Metal (m) - metal (m′) or metal (m) - chalcogen (y): (f) m=Ga, In, Tℓ, (g) m=Au, m′=Cs.

The continuous metal-nonmetal transition of an alkali metal-alkali halide solution above T_c is illustrated for the case of K-KBr in Fig. 2 (11). The DC electrical conductivity varies by four orders of magnitude between that of pure K metal and the ionic conductivity of KBr. In the dilute metal range, the conductivity increases rapidly with increasing metal concentration. The Cs-CsCl system, which has no phase separation, exhibits conductivity behavior similar to K-KBr above its critical temperature (12).

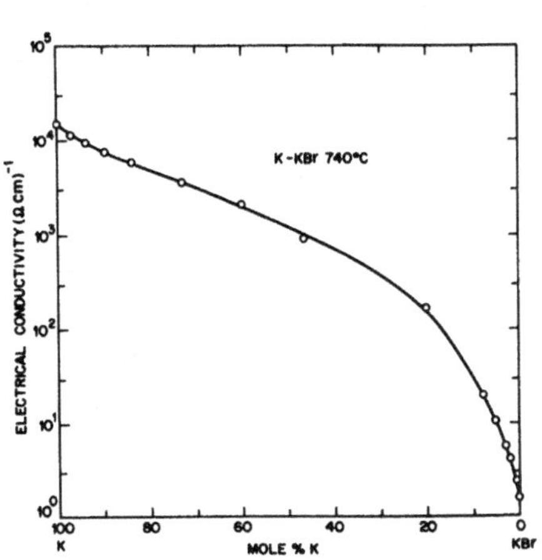

Figure 2. DC electrical conductivity versus composition in liquid K-KBr solutions at 740°C (11).

Some chalcogen alloys such as Ga_2Te_3, for which the chemical bonding may be fairly covalent, also transform to poor metals as the temperature is raised well above the melting point (3,7). Highly ionic melts like the alkali halides, in contrast, remain ionic conductors ($\sigma < 10\ \Omega^{-1}\ cm^{-1}$) over the full experimental temperature ranges investigated so far (13).

The properties of ionic liquids exhibiting metal-nonmetal transitions are reviewed in detail elsewhere (1-4,7,14). In this paper I will concentrate on the prototype systems, the alkali metal-alkali halide solutions and the ionic alloy Cs-Au. I will emphasize application of magnetic measurements, especially nuclear magnetic resonance (NMR), to investigate the following crucial questions:

(1) What is the state of the excess electrons when low concentrations of metal are added to a salt? What is the distribution in space of such states and what are their dynamic properties?

(2) How do the excess electrons interact and move as the metal concentration is increased? What is the relationship between the metal-nonmetal transition and the liquid-liquid phase separation?

I will show how the microscopic details revealed by NMR, a highly local probe, provide at least partial answers to these questions and suggest a qualitative basis for understanding the metal-nonmetal transitions in ionic liquids.

II. SURVEY OF MAGNETIC PROPERTIES

A. Magnetic Susceptibility

The static magnetic susceptibility of the solutions considered in this paper consists of three important contributions:

$$\chi_{total} = \chi^{e}_{para} + \chi^{e}_{dia} + \chi^{ion}_{dia} \tag{1}$$

where χ^{e}_{para} is the paramagnetic spin susceptibility of the excess electrons, χ^{e}_{dia} is the diamagnetic susceptibility of the excess electrons, and χ^{ion}_{dia} is the diamagnetic susceptibility of the ions. It was recognized by Nachtrieb 20 years ago that χ^{e}_{para} is a sensitive indicator of the state of excess electrons in metal-molten salt solutions (15). For example, χ^{e}_{para} vanishes if the electrons form spin-paired species, it is relatively large for weakly interacting localized spins, and is reduced by Fermi statistics if the electrons become itinerant and degenerate.

The diamagnetic total susceptibility of pure alkali halides is progressively reduced on addition of metal, an effect which is attributed to an increasing contribution from χ^{e}_{para} (12,16-19). A similar result was obtained for Cs-CsAu (20). However, quantitative evaluation of χ^{e}_{para} is difficult in these cases because all three contributions to χ_{total} are of comparable magnitude and to extract χ^{e}_{para} it is necessary to make explicit assumptions regarding the composition dependence of χ^{e}_{dia} and χ^{ion}_{dia}. The situation is somewhat more favorable for divalent and polyvalent metal-halide solutions where measurement of χ_{total} can clearly identify those cases such as $Cd\text{-}CdCl_2$ where the excess metal predominantly forms diamagnetic species (15).

B. Electron Spin Resonance (ESR)

The electron spin susceptibility χ^{e}_{para} can be determined directly from the integrated intensity of the electron spin resonance. Furthermore, dynamic properties of the electron spin can be probed using this technique by studies of the ESR linewidth or spin relaxation rates. Regrettably, experimental problems related to high temperatures and large linewidths render observation of ESR signals extremely difficult in most alkali metal-alkali halide solutions. The groups in Chicago (12,21,22) and Marburg (19) have nevertheless achieved some success with these measurements and the ESR technique holds promise for the future.

C. Nuclear Magnetic Resonance (NMR)

Nuclear magnetic resonance and other methods exploiting nuclear magnetism probe the electronic magnetization indirectly. The hyperfine interaction of the nuclei with unpaired electrons produces fluctuating local magnetic fields which can strongly affect the NMR properties (23). The time-averaged local field causes a shift of the resonant field while the fluctuations provide a mechanism for nuclear spin relaxation.

The Hamiltonian of the nuclear spin system in the presence of a hyperfine interaction can be written as the sum of the Zeeman term and an interaction term

$$H = -\gamma_n \hbar \sum_i I_z^i H_0 + \sum_{ij} A_{ij}(t)\, \vec{I}^i \cdot \vec{S}^j \tag{2}$$

where γ_n is the nuclear gyromagnetic ratio, H_0 is the applied field, $\vec{I}$ and $\vec{S}$ are the nuclear and electronic spin operators, respectively, and $A_{ij}(t)$ is the time-dependent hyperfine coupling.

The resonance shift derived from Eqn. (2) is

$$\frac{H_{ref} - H_0}{H_0} = \frac{\Delta H}{H_0} = (N_0 \gamma_e \gamma_n \hbar^2)^{-1} <A> \chi^{e}_{para} \tag{3}$$

where H_0 is the applied field at resonance, H_{ref} is the resonant field for an appropriate reference compound, N_0 is Avogadro's number, γ_e is the electronic gyromagnetic ratio, and χ^{e}_{para} is taken to be the molar electronic paramagnetic susceptibility. The average hyperfine coupling is determined by the electron probability amplitude $|\Psi(R_i)|^2$ which in a liquid is averaged over the positions $\vec{R}_i$ of all resonant nuclei:

$$<A> = <|\Psi(R_i)|^2> \tag{4}$$

In addition to the hyperfine shift described by Eqn. (3) there is, in general, a small chemical shift associated with the orbital magnetization of the ion cores. Compared with the shift due to spin paramagnetism, the chemical shifts are negligible except for very low concentrations of excess electrons.

Fluctuations of the hyperfine field can be described by an autocorrelation function

$$G_{+-}(t) = \frac{1}{4} < \sum_{ij} A_{ij}(t) S_+^j(t) A_{ij}(0) S_-^j(0) > \tag{5}$$

where, as usual, $S_{\pm} = S_x \pm iS_y$. The rate $1/T_1$ for nuclear spin-lattice relaxation is then given by the Fourier transform evaluated at the nuclear Larmor frequency ω_o:

$$^1/T_1 = \left(^2/\hbar^2\right) \int_{-\infty}^{\infty} dt\, e^{-i\omega_o t}\, G_{+-}(t)\ . \qquad (6)$$

Note that relaxation may result from either spin fluctuations $[S_{\pm}(t)]$ or fluctuations of the coupling $A_{ij}(t)$ due to the relative motion of nuclei and electrons. For the liquids of interest in this paper, the latter is assumed to be the case.

The time dependence of $G_{+-}(t)$ can usually be characterized by a correlation time τ according to

$$G_{+-}(t) = G_{+-}(0)\, e^{-t/\tau}\ . \qquad (7)$$

The correlation time for electrons in liquids is normally sufficiently short that $\gamma_e H_0 \tau << 1$. In this case, the relaxation rate is proportional to τ

$$^1/T_1 \sim n <A^2> \tau \qquad (8)$$

where n is the number of unpaired electrons in the system.

III. EXPERIMENTAL RESULTS

A. Resonance Shifts

The resonance shifts of ^{133}Cs and ^{127}I in Cs-CsI-I solutions are shown in Fig. 3 (24,25). This is a typical system exhibiting a continuous metal-nonmetal transition and no phase separation. The shift is strongly concentration dependent, decreasing from the large shift (Knight shift) of pure Cs metal to the small chemical shift of CsI. Addition of excess I to CsI produces no significant change in the resonance shift. Comparison of the relative magnitudes of the ^{133}Cs and ^{127}I shifts yields the important result that the *electronic paramagnetism is most strongly associated with the metal nuclei* in both the metallic range (60 mole % Cs) and close to the pure salt. The resonance shift of ^{133}Cs in Cs-CsAu solutions (Fig. 4) is qualitatively similar to Cs-CsI although in Cs-CsAu the increase in shift occurs much closer to the stoichiometric alloy (26,27). Because of its poor NMR properties, no data exist for ^{197}Au in Cs-CsAu.

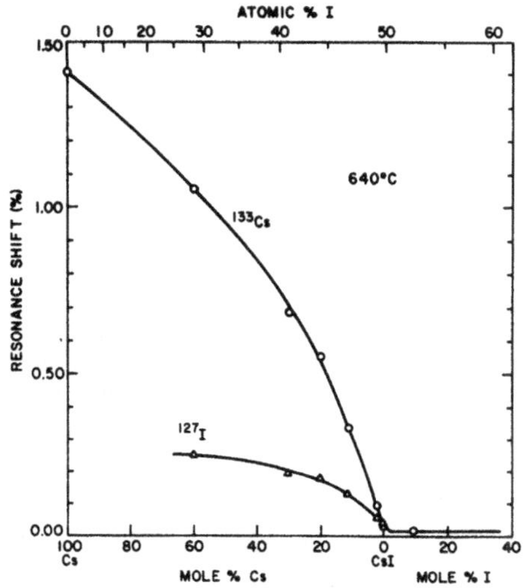

Figure 3. Resonance shifts ΔH/H versus composition for ^{133}Cs and ^{127}I in liquid Cs-CsI-I at 640°C (24,25).

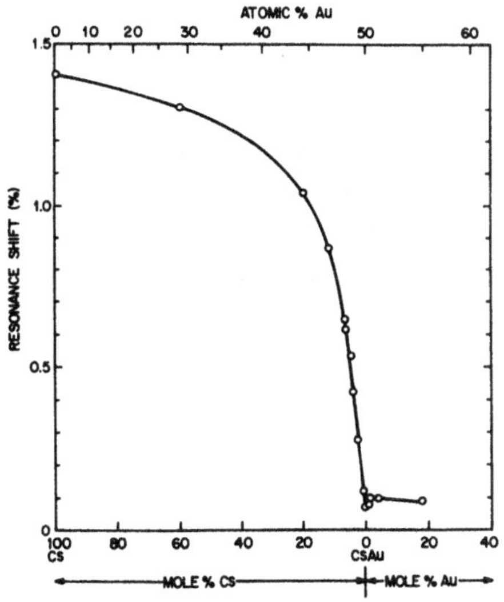

Figure 4. Resonance shift ΔH/H versus composition for ^{133}Cs in liquid Cs-CsAu-Au solutions at 600°C (26,27).

B. Nuclear Relaxation Rates

Nuclear relaxation rates for ^{133}Cs and ^{127}I in Cs-CsI-I are shown in Fig. 5 (24,25). The most striking features are the strong peaks in the relaxation rates in the dilute metal range and the extreme sensitivity of the ^{133}Cs rates to small amounts of excess metal. A sample having about 0.5 mole % excess Cs has a relaxation rate more than 5 orders of magnitude higher then pure CsI.

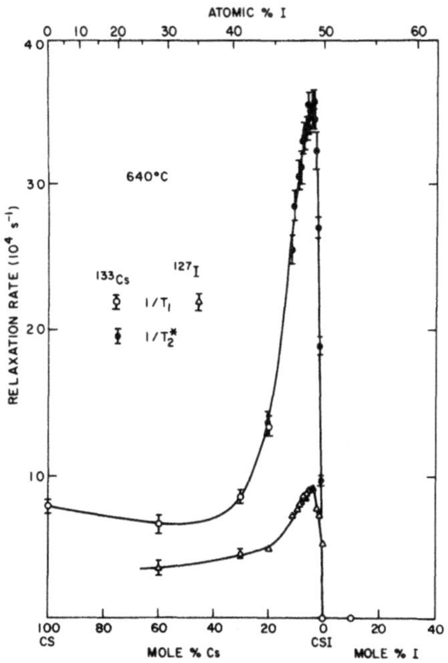

Figure 5. Nuclear spin-lattice relaxation rates ($1/T_1$), open points and spin-phase memory times ($1/T_2^*$), closed points, versus composition for ^{133}Cs and ^{127}I in liquid Cs-CsI-I solutions at 640°C (24,25).

The relatively small peak in the ^{127}I rate again indicates that the excess electrons couple more strongly to the Cs nuclei than to the I. The ^{127}I background relaxation in pure CsI is stronger than that of ^{133}Cs because of nuclear electric quadrupole relaxation. This process is ineffective for ^{133}Cs because of its small electric quadrupole moment. The ^{133}Cs relaxation rates of Cs-CsAu-Au shown in Fig. 6 (26,27) are qualitatively similar to those of Cs-CsI-I but there are some essential differences.

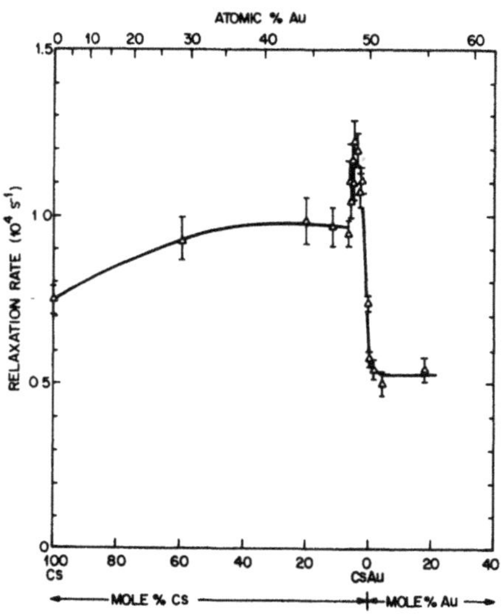

Figure 6. Nuclear relaxation rates versus composition for ^{133}Cs in liquid Cs-CsAu-Au solutions at 600°C (26,27)

The relaxation rate peak in Cs-CsAu is only 1/3 the magnitude of that for Cs-CsI and its onset is much more abrupt and closer to stoichiometry. Even very close to stoichiometric CsAu there is a substantial relaxation rate, unlike pure CsI.

IV. LOCALIZATION

Since the number of excess electrons decreases monotonically across the concentration range from metal to salt, we can conclude from Eq. 8 that the relaxation rate peaks in Cs-CsI and Cs-CsAu are due to significantly longer correlation times in the range of low metal concentration. Estimates of the correlation times using simple models show, indeed, that the electrons are strongly localized in the dilute metal limit.

Let us consider first the metallic range. Pure liquid alkali metals, and Cs in particular, can be very well represented by the nearly free electron model. In this limit, the electron mean free path λ exceeds the average interatomic separation a and transport can be described by a weak scattering approximation as in the Ziman theory (28). The hyperfine correlation time in the metallic limit is roughly $\tau_m \sim a/v_F$, i.e. the time required for an electron with the Fermi velocity v_F to move from one atom to the next. Since τ_m is independent of λ, nuclear relaxation is unrelated to the transport coefficients and is given by the well-known Korringa rate (29)

$$\left(\frac{1}{T_1}\right)_{Korr.} = \frac{4\pi kT}{\hbar}\left(\frac{\gamma_n}{\gamma_e}\right)^2\left(\frac{\Delta H}{H_o}\right)^2 \tag{9}$$

Now as anions (I^- or Au^-) are added to the metal, λ decreases until $\lambda \sim a$, the so-called Ioffe-Regel limit (30). Beyond this, as long as the electrons remain itinerant and strongly scattering, transport is diffusive (31). The nuclear relaxation rate in this range is given by the approximate expression (32)

$$\left(\frac{1}{T_1}\right) \cong \left(\frac{1}{T_1}\right)_{Korr.} \cdot \frac{\tau}{\tau_m} \tag{10}$$

where the correlation time τ represents the time a diffusing electron remains near a particular nucleus. Since τ_m may be estimated and $(^1/T_1)_{Korr.}$ is easily calculated from the measured resonance shifts using Eq. (9), Eq. (10) permits us to estimate τ from the measured rate $^1/T_1$.

If the electrons are strongly localized and obey classical statistics, the relaxation rate is

$$\frac{1}{T_1} = \frac{1}{2} c_s <(A/\hbar)^2> \tau \tag{11}$$

where c_s is the mole fraction of localized, unpaired electrons. On substituting Eq. (3) in Eq. (11) using the Curie susceptibility for χ^e_{para}, we can eliminate c_s to obtain:

$$\frac{1}{T_1} = \frac{2kT}{\hbar^2}\left(\frac{\gamma_n}{\gamma_e}\right)\frac{<A^2>}{<A>}\frac{\Delta H}{H_o}\tau \tag{12}$$

Since we can estimate $<A^2>/<A>$ from the magnitude of the shift, we can extract τ in the localized limit using Eq. (12).

The results of an analysis using these models are illustrated in Fig. 7 for Cs-CsI. The correlation time increases dramatically from a typically metallic value of $\sim 10^{-15}$s in Cs to $\sim 10^{-12}$s in the dilute metal. In the metal-nonmetal transition range there is rough agreement between the results of the itinerant and localized models. It is significant that the correlation times for dilute excess metal in CsI (and similarly in Cs-CsCl and Na-NaBr) are comparable with the structural lifetime associated with ionic diffusion, i.e. roughly 10^{-12}s. I will make use of this fact shortly.

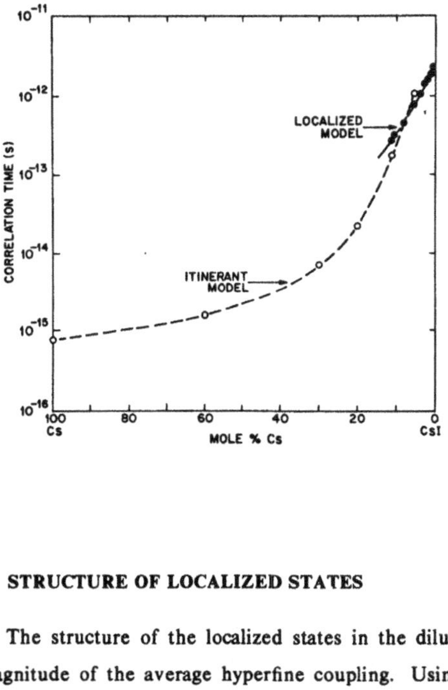

Figure 7. Hyperfine field correlation time τ versus composition for liquid Cs-CsI solutions at 640°C (24). Correlation times for the itinerant and localized models are derived using Eqns. (10) and (12), respectively.

V. STRUCTURE OF LOCALIZED STATES

The structure of the localized states in the dilute metal limit can be investigated by consideration of the magnitude of the average hyperfine coupling. Using the Curie law to calculate χ^e_{para}, we can estimate $<A>$ from the magnitude of $\Delta H/H_0$ using Eq. (3). We illustrate for the case of Na-NaBr for which experiment yields

$$<|\Psi(Na)|^2> = 1.5 \times 10^{24} \text{ cm}^{-3}$$

This value is substantially smaller than the free atom value $|\Psi(Na)|^2_{atom} = 5.1 \times 10^{24}$ cm^{-3} (33) indicating clearly that the localized states do not resemble those of the neutral Na atom.

An alternative to atomic states is provided by the 'F-center' model suggested by Pitzer (34). Here one supposes that the electron enters an anion vacancy, i.e. a cavity coordinated mainly by cations. Such sites may be seen, for example, in Woodcock's model, constructed on the basis of molecular dynamics calculations (35). The optical absorption spectra of dilute excess electrons in molten alkali halides are similar in most respects to those of the analogous crystalline F-centers (36-39). The energy levels of the 'F-center' in liquid alkali halides have been calculated by Senatore, Parrinello and Tosi (40) using a model which is discussed by Tosi elsewhere in these Proceedings.

An estimate of the expected average hyperfine coupling for F-centers is provided by ENDOR experiments in alkali halide crystals. Such measurements determine directly the hyperfine field on the near-neighbor and more distant ion shells. For F-centers in crystalline NaBr, $<|\Psi(\mathrm{Na})|^2> \simeq 1.8 \times 10^{24}\ \mathrm{cm}^3$ (41), a value within 20% of the liquid-state value. A similar correspondence was found for the Cs-halides (25). Further support for the 'F-center' model is provided by the limiting values of $\tau \sim 10^{-12}$s. In contrast to neutral atoms or other paramagnetic species, one should expect the lifetime of the 'F-center' site to be limited by the ionic diffusion times as is, in fact, observed.

VI. EFFECT OF INCREASING METAL CONCENTRATION - METAL-NONMETAL TRANSITION AND PHASE SEPARATION

Beginning from a model of non-interacting localized 'F-centers' at dilute metal concentrations in the salt, we must now consider what happens as the metal concentration is increased. I shall discuss this in the context of an old dilemma illustrated by the data shown in Fig. 8. The concentration dependence of the electrical conductivity σ of Na-NaBr solutions (42) (and other Na-Na halide solutions) differs strikingly from that of solutions based on heavier alkali metals such K-KBr (11) or Cs-CsCl (12). A plot of σ versus concentration exhibits negative curvature for the Na systems, corresponding to a decreasing equivalent molar conductance, while for the heavier alkali systems, positive curvature or increasing equivalent conductance is observed. An important fact which has not been adequately acknowledged in previous discussions of this effect, is that at the temperatures of measurement, just above the melting point of the salts, the Na-NaX systems have very limited solubility of metal. That is, the data were taken far below T_c where there is no continuous metal-nonmetal transition. For Cs-CsCl or the K-KX systems, the data were obtained in the range of complete miscibility or just below T_c where there is a wide range of solubility.

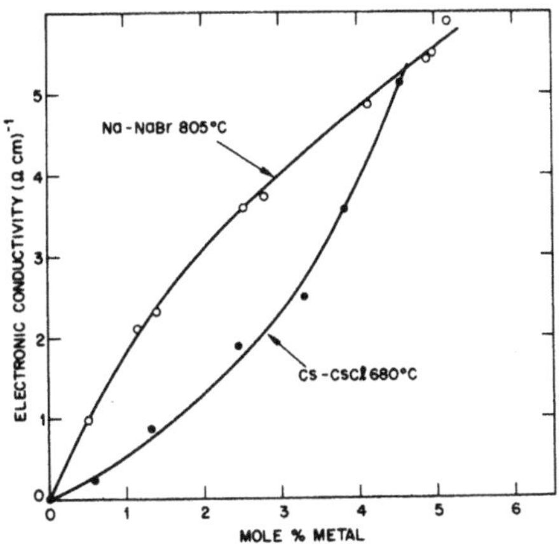

Figure 8. Electronic conductivity versus composition in dilute metal range for Na-NaBr at 805°C (42) and Cs-CsCl at 680°C (12). Electronic contribution was obtained by subtracting the conductivity of the pure salts from the total conductivity of the solutions.

It was suggested by Bronstein and Bredig (1,42) that the anomalous behavior of Na-NaX systems might be due to formation of an increasing proportion of diatomic species Na_2 which reduce the number of conducting electrons at higher concentration. The fact that the anomalous negative curvature is reduced at higher temperatures is consistent with this notion as is the absence of the effect for the heavier metals whose dimers are less stable in the vapor (1). In models introduced later by Katz and Rice (43) and Durham and Greenwood (44) it was proposed that the mobility of the excess electrons in Na-NaX solutions decreases with added metal concentration due to increased disorder (45) and screening effects. Recent NMR studies of Na-NaBr (46) have shown that his is incorrect and that, in fact, the 'anomalous' behavior of Na-NaBr is due solely to a relative decline in the number conducting electrons as implied by Bronstein and Bredig's original proposal.

In the localized range and at concentrations low enough that $\langle A^2\rangle/\langle A\rangle$ remains constant, Eq. (12) may be used to extract the concentration dependency of τ from measurements of $1/T_1$ and $\Delta H/H_0$. For a diffusive transport mechanism, the mobility $\mu \; \alpha \; \tau^{-1}$ so this analysis yields also the variation of the mobility. The relative concentration dependences of μ determined from NMR in Na-NaBr (46) and Cs-CsX (25) solutions are shown in Fig. 9. The electronic mobility of Cs-CsCl, extracted from the conductivity by assuming all excess electrons conduct, is also shown. It is found that mobility in Na-NaBr is unchanged between 1 and 4 mole % excess metal while that of Cs-CsI roughly doubles over the same range. Furthermore, the relaxation rates in Na-NaBr are accurately proportional to the electronic part of the conductivity showing that at higher metal concentrations, electrons are removed by formation of states which are both non-magnetic (i.e. spin-paired) and do not contribute to the electronic conductivity. In Cs-CsX solutions, in contrast, addition of metal increases both the number of conducting electrons and their mobility.

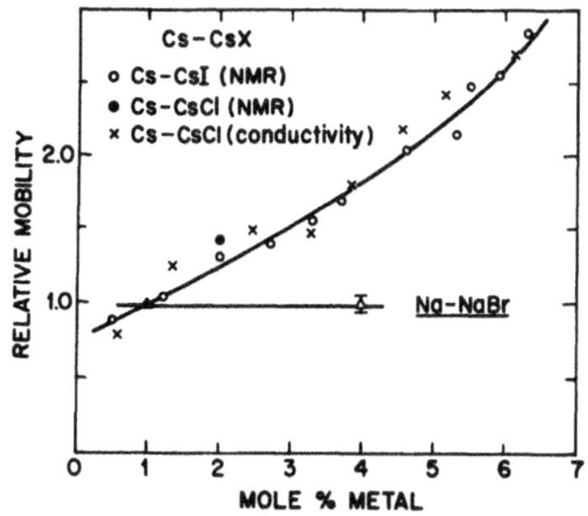

Figure 9. Relative electronic mobility from Eqn. 10 versus composition in the dilute metal range for Cs-CsI, Cs-CsCl, and Na-NaBr. Mobility of Cs-CsCl from conductivity (12) was derived assuming all excess electrons are equivalent and conducting.

The foregoing analysis leads me to the following speculation concerning the role of polyatomic species in the metal-nonmetal transition and phase separation. We suppose, as the evidence indicates, that at dilute metal concentrations the electrons enter localized states resembling F-centers. The rate of transport in this limit is determined by the lifetime of the ionic configuration ($\sim 10^{-12}$s). When the ions diffuse away, destroying a favorable site for localization, the electron moves to a new site, typically about 20 Å away according to the observed conductivity. Now at higher metal concentrations, the possibility of forming M_2 dimers (and other 'clusters') increases. If the dimer is relatively stable, that is, if its lifetime is long compared with the ionic diffusion time, then the mobility of the remaining electrons should be independent of concentration. When a sufficient concentration of dimers is reached, the second, metallic phase is formed. This is the case for the Na-NaX systems well below T_c. Above T_c and in the Cs-CsX systems, if the dimer lifetime is shorter than the ionic diffusion time, there should be rapid equilibrium between the 'F-centers' and dimers

$$2M^+ + 2e^- \rightleftarrows M_2$$

This rapid formation and dissociation of the dimer enhances the mobility, since the electrons may localize at new sites after dissociation, thus moving faster than the rate determined by ionic diffusion. This mechanism not only provides a strongly concentration-dependent mobility, but places an increasing proportion of the conducting electrons on the ions (in dimers, etc.) as required for the eventual transition to a metallic state. A similar scheme, also based on the sequence F-centers → spin-paired species → "clusters", was proposed recently by Nicoloso (19). According to the present picture, it is the stability of the polyatomic, spin-paired species which governs the liquid-liquid phase transition and the continuous metal-nonmetal transition. An analogy between these polyatomic species in metal-molten salt solutions and the charged polyatomic 'droplets' discussed at this conference by Hensel for Hg vapor is a fascinating possibility which deserves further attention.

References

1. M. A. Bredig, in *Molten Salt Chemistry*, ed. M. Blander (Interscience, New York, 1964) p. 367.
2. N. F. Mott, *Metal-Insulator Transitions* (Taylor and Francis, London, 1974).
3. M. Cutler, *Liquid Semiconductors* (Academic Press, New York, 1977).
4. W. W. Warren, Jr. in *Advances in Molten Salt Chemistry*, Vol. 4, ed. G. Mamantov and J. Braunstein (Plenum, New York, 1981) p. 1.
5. H. Yokokawa, O. J. Kleppa, and N. H. Nachtrieb, J. Chem. Phys. *71*, 4099 (1979).
6. H. Yokokawa and O. J. Kleppa, J. Chem. Phys. *76*, 5574 (1982).
7. V. M. Glazov, S. N. Chizhevskaya, and N. N. Glagoleva, *Liquid Semiconductors* (Plenum, New York, 1969).
8. M. Hansen nd K. Anderko, *Constitution of Binary Alloys* (McGraw-Hill, New York, 1959) 2nd edition.
9. R. P. Elliott, *Constitution of Binary Alloys, First Supplement* (McGraw-Hill, New York, 1965).
10. F. A. Shunk, *Constitution of Binary Alloys, Second Supplement* (McGraw-Hill, New York, 1969).
11. H. R. Bronstein, A. S. Dworkin, and M. A. Bredig, J. Chem. Phys. *31*, 677 (1962).
12. N. H. Nachtrieb, C. Hsu, M. Sosis, and P. A. Bertrand, in *Proceedings of the International Symposium on Molten Salts*, ed. J. P. Pemsler, J. Braunstein, and K. Nobe (Electrochemical Society, Princeton, 1976) p. 506.
13. A. Klemm, in *Molten Salt Chemistry*, ed. M. Blander (Interscience, New York, 1964) p. 535.
14. J. D. Corbett, in *Fused Salts*, ed. B. Sundheim (McGraw-Hill, New York, 1964) Chap. 6.
15. N. H. Nachtrieb, J. Phys. Chem. *66*, 1163 (1962).
16. M. Bettman, J. Chem. Phys. *44*, 3254 (1966).
17. R. H. Arendt and N. H. Nachtrieb, J. Chem. Phys. *53*, 3085 (1970).
18. G. Steinleitner, *Dissertation* (U. Marburg, 1978).
19. N. Nicoloso, *Dissertation* (U. Marburg, 1982).
20. W. Freyland and G. Steinleitner, Ber. Bunsenges, Phys. Chem. *80*, 810 (1976).
21. M. Sosis, *Thesis* (J. Franck Institute, U. Chicago, 1974).
22. P. A. Bertrand, *Thesis* (J. Franck Institute, U. Chicago, 1978).
23. See, for example, A. Abragam, *The Principles of Nuclear Magnetism* (Oxford, London, 1961)
24. S. Sotier and W. W. Warren, Jr., J. Physique *41*, C8-40 (1980).
25. W. W. Warren, Jr. and S. Sotier, in *Proc. Third Int. Symposium on Molten Salts*, ed. G. Mamantov, M. Blander, and G. P. Smith (The Electrochemical Society, Princeton, 1981) p. 95.
26. R. Dupree, D. Kirby, W. Freyland, and W. W. Warren, Jr. Phys. Rev. Lett. *45*, 130 (1980)
27. R. Dupree, D. J. Kirby, W. Freyland, and W. W. Warren, Jr., J. Physique *41*, C8-16 (1980)
28. J. M. Ziman, Phil Mag. *6*, 1013 (1961).
29. J. Korringa, Physica *16*, 601 (1950).
30. A. F. Ioffe and A. R. Regel, Progr. Semicond. *4*, 239 (1960).

31. M. H. Cohen, J. Non-Cryst. Solids *4*, 391 (1970).
32. W. W. Warren, Jr., Phys. Rev. B*3*, 3708 (1971).
33. P. Kusch and H. Taub, Phys. Rev. *75*, 1477 (1949).
34. K. S. Pitzer, J. Am. Chem. Soc. *84*, 2025 (1962).
35. L. V. Woodcock, Nature (Phys. Sci) *232*, 63 (1971).
36. E. Mollwo, Nachr. Gesell. Wiss. Gottingen, Math-Physik K., Fachgruppe II, 1:203 (1935).
37. D. M. Gruen, M. Krupelt, and I. Johnson, *in Molten Salts, Characterization and Analysis*, ed. G. Mamantov (Dekker, New York, 1969) p. 169.
38. J. F. Rounsaville and J. J. Lagowski, J. Phys. Chem. *72*, 1111 (1968).
39. H. -J. Yuh, *Thesis* (U. Chicago, 1981).
40. G. Senatore, M. Parrinello, and M. Tosi, Phil, Mag. *41*, 595 (1980).
41. H. Seidel and H. C. Wolf, in *Physics of Color Centers*, ed. W. B. Fowler (Academic Press, New York, 1964) p. 109.
42. H. R. Bronstein and M. A. Bredig, J. Am. Chem. Soc. *80*, 2077 (1958).
43. I. Katz and S. A. Rice, J. Am. Chem. Soc. *94*, 4824 (1972).
44. P. J. Durham and D. A. Greenwood, Phil Mag. *33*, 427 (1976).
45. P. W. Anderson, Phys. Rev. *109*, 1492 (1958).
46. S. Sotier and W. W. Warren, Jr. (unpublished).

COLLECTIVE DYNAMICS AND TRANSPORT IN BINARY IONIC MIXTURES

J.P. HANSEN, B. BRAMI and F. JOLY

Laboratoire de Physique Théorique des Liquides*,
Université Pierre et Marie Curie ; 75230 Paris Cedex 05

ABSTRACT.

Recent Molecular Dynamics results for the charge fluctuation spectra of binary ionic mixtures immersed in a neutralizing background are analysed on the basis of the coupled generalized Langevin equations for the partial densities. Good agreement with the simulation data is achieved, except at low frequencies. The spectrum of the k-dependent interdiffusion current correlation function is shown to exhibit well separated low and high frequency peaks linked to mutual diffusion of the two species and to plasma oscillations respectively

1. INTRODUCTION.

It has recently become clear that the simple "one component plasma" (OCP) model offers an excellent starting point for the investigation of thermodynamic and structural properties of the liquid alkali metals [1,2]. The OCP is a system of point ions immersed in a uniform, neutralizing background provided by the degenerate Fermi gas of conduction electrons. The reduced equilibrium properties of the classical ionic component depend on a single Coulomb coupling parameter [3] :

$$\Gamma = \frac{Z^2 e^2}{a k_B T} \tag{1.1}$$

where Z is the ionic valence, e the proton charge and a is the "ion-sphere radius" which is defined in terms of the number-density, $n = N/\Omega$, by :

$$a = \left(\frac{3}{4\pi n}\right)^{1/3} \tag{1.2}$$

* Equipe associée au C.N.R.S.

The close relation between the OCP and simple liquid metals is illustrated by the striking similarity between the static structure factors of the OCP and of the alkali just above melting [1] ; the agreement breaks down only at small wavenumbers, due to electron screening, which is ignored in the OCP model [2,4].

The present lecture is devoted to the binary extension of the OCP, the so-called "binary ionic mixture" (BIM). This model can be considered as the first, admittedly crude, step towards a description of binary alloys with simple band structures. The BIM is also an excellent model for the study of very dense ionic matter in planetary interiors [5] and in degenerate stars. In fact the work presented in this paper was originally motivated by some astrophysical problems concerning ionic self-diffusion and inter-diffusion in white dwarfs and at the surface of neutron stars.

2. CHARACTERISTIC PARAMETERS OF THE MODEL.

We consider a binary mixture containing N_1 ions of charge Z_1e and mass M_1 and N_2 ions of charge Z_2e and mass M_2, immersed in a rigid, uniform background of opposite charge. The respective number densities and concentrations are $n_\alpha = N_\alpha/\Omega$ and $x_\alpha = N_\alpha/N$ ($N = N_1+N_2$; $\alpha = 1,2$). We denote by $\overline{Z} = x_1Z_1+x_2Z_2$ and by $\overline{M} = x_1M_1+x_2M_2$ the mean valence and the mean mass of the ions. It will be convenient to introduce the relative valences and masses $z_\alpha = Z_\alpha/\overline{Z}$ and $m_\alpha = M_\alpha/\overline{M}$ respectively. The total potential energy of a periodic system in a volume Ω can be cast in the form :

$$V_N(\Omega) = \frac{1}{2\Omega} \sum_{\vec{k}}{}' \frac{4\pi e^2}{k^2} \left[\rho_{\vec{k}Z} \, \rho^*_{\vec{k}Z} - N\overline{Z^2} \right] \tag{2.1}$$

where the prime in the summation indicates that the $\vec{k} = \vec{0}$ term has to be omitted to take proper account of the uniform background, and $\rho_{\vec{k}Z}$ denotes a Fourier component of the microscopic charge density :

$$\rho_{\vec{k}Z} = Z_1 \, \rho_{\vec{k}1} + Z_2 \, \rho_{\vec{k}2} \tag{2.2a}$$

$$\rho_{\vec{k}\alpha} = \sum_{i=1}^{N_\alpha} e^{-i\vec{k}.\vec{r}_{i\alpha}} \quad ; \quad \alpha = 1,2 \tag{2.2b}$$

A thermodynamic state of the classical mixture can be characterized by the coupling constant Γ defined by eq.(1.1) (with Z taken conventionally to be the smaller of the two valences Z_α) and by the concentration x_1 ($x_2 = 1-x_1$). The static properties of the model are well documented [7].

The time scale of the microscopic dynamics is set by the plasma frequency. But contrarily to the case of the OCP, there is some ambiguity in defining this frequency [8,9]. Mean field theory leads naturally to "kinetic" plasma frequency :

$$\Omega_p^2 = \Omega_{p1}^2 + \Omega_{p2}^2 \tag{2.3a}$$

$$\Omega_{p\alpha} = \left[\frac{4\pi n_\alpha Z_\alpha^2 e^2}{M_\alpha}\right]^{1/2} \tag{2.3b}$$

while a continuum description would lead to the "hydrodynamic" frequency :

$$\omega_p = \left[\frac{4\pi n \bar{Z}^2 e^2}{\bar{M}}\right]^{1/2} \tag{2.4}$$

The two frequencies differ generally by a small, but significant amount :

$$\delta^2 = \Omega_p^2 - \omega_p^2 = \omega_p^2 x_1 x_2 m_1 m_2 \left[\frac{z_1}{m_1} - \frac{z_2}{m_2}\right]^2 \geqslant 0 \tag{2.5}$$

For an equimolar H^+-He^{++} alloy for instance, the relative difference $\delta/\omega_p \simeq 5\%$. Note that $\Omega_p \equiv \omega_p$ either if x_1 (or x_2) = 0 (OCP), or if the condition :

$$\frac{Z_1}{Z_2} = \frac{M_1}{M_2} \tag{2.6}$$

is satisfied. The latter situation is particularly interesting, and will be re-examined in more detail. If the condition (2.6) is not fulfilled, the long wavelength frequency of the plasmon mode must go to Ω_p in the weak coupling limit ($\Gamma \to 0$) and is expected to vary with temperature [8,9].

3. CHARGE, MASS AND CONCENTRATION FLUCTUATION SPECTRA.

The fundamental dynamical variables for a study of charge, mass and concentration fluctuations in the BIM are the time-dependent Fourier components (2.2b) of the microscopic partial densities and the associated longitudinal currents :

$$j^{\ell}_{\vec{k}\alpha}(t) = \frac{1}{\omega_{o\alpha}(k)} \sum_{i=1}^{N_\alpha} \vec{k}.\vec{v}_{i\alpha}(t)\, e^{-i\vec{k}.\vec{r}_{i\alpha}(t)} \tag{3.1}$$

where $\omega_{o\alpha}(k) = v_{o\alpha}\, k = (k_B T/m_\alpha)^{1/2} k$

The two sets of dynamical variables are related, by the continuity equations (where the dot represents a time derivative) :

$$\dot{\rho}_{\vec{k}\alpha}(t) = -i\, \omega_{o\alpha}(k)\, j^{\ell}_{\vec{k}\alpha}(t) \tag{3.2}$$

We shall concentrate on the density and current correlation functions :

$$F_{\alpha\beta}(k,t) = \frac{1}{(N_\alpha N_\beta)^{1/2}} \langle \rho_{\vec{k}\alpha}(t)\, \rho^{*}_{\vec{k}\beta}(0) \rangle \tag{3.3a}$$

$$C_{\alpha\beta}(k,t) = \frac{1}{(N_\alpha N_\beta)^{1/2}} \langle j^{l}_{\vec{k}\alpha}(t)\, j^{l*}_{\vec{k}\beta}(0) \rangle$$

$$= -\frac{1}{\omega_{0\alpha}(k)\,\omega_{0\beta}(k)}\, \ddot{F}_{\alpha\beta}(k,t) \tag{3.3b}$$

and on the partial dynamical structure factors and the current fluctuation spectra :

$$S_{\alpha\beta}(k,\omega) = \frac{1}{2\pi} \int_{-\infty}^{+\infty} e^{i\omega t}\, F_{\alpha\beta}(k,t)\, dt \tag{3.4a}$$

$$\hat{C}_{\alpha\beta}(k,\omega) = \frac{\omega^2}{\omega_{0\alpha}(k)\,\omega_{0\beta}(k)}\, S_{\alpha\beta}(k,\omega) \tag{3.4b}$$

Note that :

$$F_{\alpha\beta}(k, t=0) = S_{\alpha\beta}(k) \tag{3.5a}$$

$$C_{\alpha\beta}(k, t=0) = \delta_{\alpha\beta} \tag{3.5b}$$

where the $S_{\alpha\beta}(k)$ are the partial static structure factors. The physically important linear combinations of the partial densities are the charge and mass densities and the local concentration variable

$$\rho_{\vec{k}Z}(t) = Z_1\, \rho_{\vec{k}1}(t) + Z_2\, \rho_{\vec{k}2}(t) \tag{3.6a}$$

$$\rho_{\vec{k}M}(t) = M_1\, \rho_{\vec{k}1}(t) + M_2\, \rho_{\vec{k}2}(t) \tag{3.6b}$$

$$\rho_{\vec{k}C}(t) = x_2\, \rho_{\vec{k}1}(t) - x_1\, \rho_{\vec{k}2}(t) \tag{3.6c}$$

Similar linear combinations hold for the associated currents. The corresponding correlation functions are straightforward combinations of the partial correlation functions (3.3). The three variables (3.6) are clearly not independent, and we find in particular that :

$$\rho_{\vec{k}Z}(t) = \frac{\bar{Z}}{\bar{M}} \left[\rho_{\vec{k}M} - (z_1 m_2 - z_2 m_1)\, \rho_{\vec{k}C} \right] \tag{3.7}$$

The ionic conductivity and the mutual diffusion constant are directly expressible in terms of the $\vec{k} = \vec{0}$ currents :

$$\vec{j}_Z(t) = Z_1\, \vec{j}_1(t) + Z_2\, \vec{j}_2(t) \tag{3.8a}$$

$$\vec{j}_C(t) = x_2\, \vec{j}_1(t) - x_1\, \vec{j}_2(t) \tag{3.8b}$$

$$\vec{j}_M(t) = M_1 \vec{j}_1(t) + M_2 \vec{j}_2(t) \equiv \vec{j}_M(0) \qquad (3.8c)$$

where the last equality follows from the conservation of total momentum. The ionic d.c. conductivity is given by the standard Kubo formula :

$$\sigma = \frac{e^2}{3\Omega k_B T} \int_0^\infty < \vec{j}'_Z(t) \cdot \vec{j}'_Z(0) > dt \qquad (3.9)$$

where $\vec{j}'_Z(t)$ is the fluctuating electric current relative to the center-of-mass motion :

$$\begin{aligned} \vec{j}'_Z(t) &= \vec{j}_Z(t) - \frac{\bar{Z}}{\bar{M}} \vec{j}_M(t) \\ &\equiv \bar{Z}\,(z_1 m_2 - z_2 m_1)\, \vec{j}_c(t) \end{aligned} \qquad (3.10)$$

The mutual diffusion constant is similarily given by :

$$D_{12} = \gamma \frac{1}{3N} \int_0^\infty < \vec{j}_c(t) \cdot \vec{j}_c(0) > dt \qquad (3.11)$$

where :

$$\gamma = \left(\frac{\partial^2 G/N k_B T}{\partial x_1^2} \right)_{P,T,N}$$

and G denotes the Gibbs free energy of the mixture ; many alloys are nearly ideal, so that : $\gamma \simeq (x_1 x_2)^{-1}$.

In view of eq.(3.10) it is immediately clear that σ and D_{12} must be proportional. Comparing eqs.(3.9-11) we find :

$$\begin{aligned} \sigma &= \frac{\omega_p^2}{4\pi} \frac{1}{\gamma \bar{v}_0^2} (z_1 m_2 - z_2 m_1)^2 D_{12} \\ &\simeq \frac{\omega_p^2}{4\pi} \frac{x_1 x_2}{\bar{v}_0^2} (z_1 m_2 - z_2 m_1)^2 D_{12} \end{aligned} \qquad (3.12)$$

where $\bar{v}_0 = (k_B T/\bar{M})^{1/2}$.

4. SOME PECULARITIES OF THE BIM.

Formula (3.12) shows that the d.c. conductivity of the BIM vanishes whenever x_1 (or x_2) = 0 (the BIM reduces then to the OCP) or when condition (2.6) is fulfilled. In either case charge and mass density fluctuations are proportional (cf. eq.(3.7)) and the vanishing of the conductivity is then a simple consequence of the conservation of total momentum. Note that whenever $\sigma = 0$, the two plasma frequencies Ω_p and ω_p coincide. The vanishing conductivity leads also to an undamped plasmon mode in the limit $k\to 0$, due to the absence of Ohmic dissipation.

This peculiar behaviour of the BIM can be understood by examining the equation of motion of any one ion :

$$M_\alpha \ddot{\vec{r}}_{i\alpha} = -\frac{\partial V_N}{\partial \vec{r}_{i\alpha}}$$

$$= Z_\alpha \mathcal{R} \frac{1}{\Omega} \sum_{\vec{k}}{}' i\vec{k} \frac{4\pi e^2}{k^2} \rho_{\vec{k}Z} e^{i\vec{k}.\vec{r}_{i\alpha}} \qquad (4.1)$$

Hence :

$$\ddot{\vec{r}}_{i\alpha} = \frac{Z_\alpha}{M_\alpha} \vec{F} \qquad (4.2)$$

where $\vec{F}$ is a force which is independent of particle species. If condition (2.6) is satisfied, we see that the acceleration of an ion for a given configuration will be independent of particle species.

Another interesting situation is that where $Z_1=Z_2$, but $M_1\neq M_2$. In that case the statics of the BIM reduce to that of the OCP, but the dynamics correspond to those of an isotopic mixture. This case can be treated by perturbation theory, in powers of the mass difference [10]. Work along these lines is in progress.

5. MOLECULAR DYNAMICS RESULTS.

We briefly summarize the most salient results of recent "molecular dynamics" (MD) computer simulations of H^+- He^{++} ($Z_2/Z_1 = 2$, $M_2/M_1 = 4$) and D^+- He^{++} ($Z_2/Z_1 = M_2/M_1 = 2$) mixtures [8]. All simulations were done at equal concentrations ($x_1= x_2 = 1/2$) and for coupling constants Γ = 0.4, 25, 40. and 62. Extensive results were obtained for the velocity autocorrelation functions $C_\alpha(t)$ of both species and for the charge density correlation function $F_{ZZ}(k,t)$ and the corresponding dynamical structure factor $S_{ZZ}(k,\omega)$. Some of the most relevant qualitative findings were the following :

a) For strong coupling ($\Gamma \gg 1$), the two $C_\alpha(t)$ oscillate in time after a fast initial decay ; although the oscillations of C_1 and C_2 are out of phase, their angular frequency is roughly equal to the plasma frequency, as in the case of the OCP [11]. The resulting self diffusion constants D_1 and D_2 drop rapidly with increasing Γ, but their ratio D_1/D_2 is roughly constant and equal to 1.7 in the strong coupling limit.

b) In the intermediate coupling range (Γ = 0.4), it was found that the charge fluctuation spectrum $S_{ZZ}(k,\omega)$ exhibits a plasmon peak which broadens rapidly with increasing wave-number due to Landau and collisional damping, while the frequency of the plasmon mode increases, in qualitative agreement with mean field theory. A dramatic isotopic effect occurs in going to the $D^+ - He^{++}$ mixture, where condition (2.6) is satisfied ; the absence of ionic conduction leads then to a considerable sharpening of the plasmon peak for a given wavenumber, as illustrated in fig. 1.

c) In the strong coupling regime (Γ = 25, 40 and 62) the charge fluctuation spectra look qualitatively very different from their intermediate coupling counterparts. Landau damping is now completely negligible since the reduced Debye wavenumber $q_D = a\, k_D \gg 1$. $S_{ZZ}(q,\omega)$ exhibits, for $q = ak \lesssim 1$, both a very sharp central peak (which is absent at intermediate couplings) and a sharp plasmon peak corresponding to weakly collision-damped plasma oscillations. While the amplitude $\sigma_{ZZ}(q,\omega = 0) = S_{ZZ}(q,\omega = 0)/S_{ZZ}(q)$ of the central peak increases sharply with Γ, the rest of the normalized spectrum $\sigma_{zz}(q,\omega)$, and in particular the plasmon peak, appears to be independent of Γ in the strong coupling regime. The plasmon mode frequency <u>decreases</u> with increasing wavenumber ; this negative dispersion for strong couplings is reminiscent of a similar behaviour in the OCP [11]. The extrapolation of the plasmon frequency to zero wavenumber yields $\omega(k = 0) \simeq 1.07\omega_p \simeq 1.02\Omega_p$, independent of coupling if $\Gamma \gg 1$.

6. A SIMPLE MEMORY FUNCTION ANALYSIS.

We have analysed the MD results for the charge fluctuation spectra by applying the Mori-Zwanzig memory function formalism to the 2x2 matrix of longitudinal current correlation functions (3.3b). The Laplace transform of this matrix :

$$\underline{\tilde{C}}(k,z) = \int_0^\infty e^{izt}\, \underline{C}(k,t)\, dt \quad ; \quad \mathrm{Im}\, z > 0 \tag{6.1}$$

satisfies the generalized Langevin equation :

$$\left\{ -iz\underline{I} + \frac{i}{z}\underline{W}(k) + \underline{\tilde{M}}(k,z) \right\} \underline{\tilde{C}}(k,z) = \underline{C}(k,t=0) = \underline{I} \tag{6.2}$$

In eq. (6.2) $\underline{I}$ denotes the unit matrix, $\underline{\tilde{M}}(k,z)$ is the Laplace transform of the memory function matrix and the frequency matrix $\underline{W}(k)$ has elements :

$$W_{\alpha\beta}(k) = \omega_{0\alpha}(k)\, \omega_{0\beta}(k) \left[\underline{S}(k)^{-1} \right]_{\alpha\beta} \tag{6.3}$$

where $\underline{S}(k)$ is the matrix of static structure factors (3.5a).

The short time expansion of the memory functions $M_{\alpha\beta}(k,t)$ and the corresponding high frequency expansion of $\tilde{M}_{\alpha\beta}(k,z)$ can be easily derived from the analoguous expan-

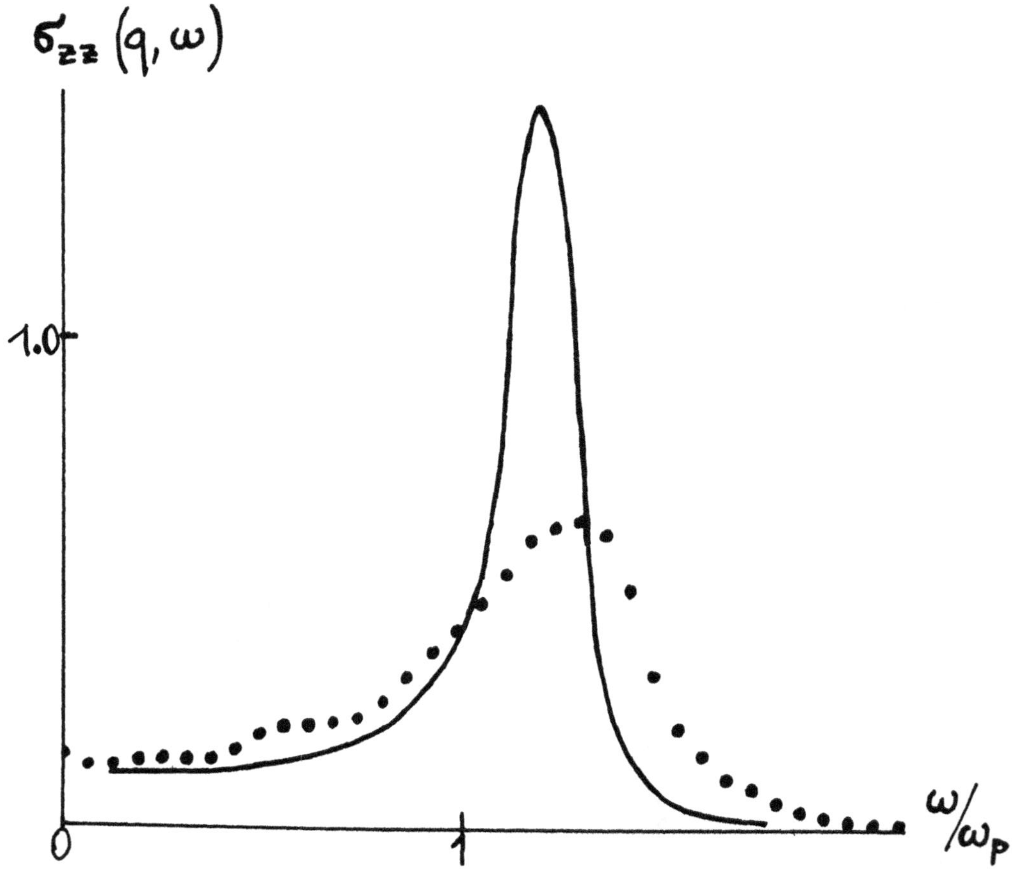

Fig. 1 : Normalized charge dynamical structure factor $\sigma_{ZZ}(q,\omega)$ as a function of ω/ω_p at $q = 0.62$ and $x_1 = x_2 = 0.5$; dots : H^+-He^{++} mixture ; full curve : D^+-He^{++} mixture. MD results from ref.[8].

sions for the $C_{\alpha\beta}(k,t)$ or, equivalently, for the $F_{\alpha\beta}(k,t)$; if the latter is written as :

$$\tilde{F}_{\alpha\beta}(k,z) = \frac{i}{z} \sum_{n=0}^{\infty} \frac{\Omega^{(2n)}_{\alpha\beta}(k)}{z^{2n}} \qquad (6.4)$$

we find that :

$$M_{\alpha\beta}(k,t=0) \equiv \mu_{\alpha\beta}(k) = \frac{\Omega^{(4)}_{\alpha\beta}(k)}{\omega_{o\alpha}(k)\,\omega_{o\beta}(k)} - W_{\alpha\beta}(k) \qquad (6.5)$$

The expressions for the frequency moments $\Omega^{(2n)}_{\alpha\beta}$ (n = 0, 1 and 2) are given in ref. [8] and can be calculated from the partial pair distribution functions which are accurately evaluated in the "hyper-netted chain" (HNC) approximation [7].

The matrix (6.5) has in general non-zero off-diagonal elements. We now follow a procedure proposed by Abramo et al. [12] and choose instead of the fundamental dynamical variables $j^{\ell}_{\vec{k}1}(t)$ and $j^{\ell}_{\vec{k}2}(t)$ their two linear combinations which diagonalize the matrix $\mu_{\alpha\beta}(k)$ for each value of k. The transformation is of the form :

$$\begin{pmatrix} \frac{1}{\sqrt{N}}\, j^{\ell}_{\vec{k}A} \\ \frac{1}{\sqrt{N}}\, j^{\ell}_{\vec{k}B} \end{pmatrix} = \begin{pmatrix} \cos\theta(k) & -\sin\theta(k) \\ \sin\theta(k) & \cos\theta(k) \end{pmatrix} \begin{pmatrix} \frac{1}{\sqrt{N_1}}\, j^{\ell}_{\vec{k}1} \\ \frac{1}{\sqrt{N_2}}\, j^{\ell}_{\vec{k}2} \end{pmatrix} \qquad (6.6)$$

The angle $\theta(k)$ in the unitary transformation matrix $\underline{T}$ is determined by the condition :

$$\mu'_{12}(k) = \left[\underline{T}\; \underline{\mu}(k)\; \underline{T}^{+}\right]_{12} = 0 \qquad (6.7)$$

In the limit $k \to 0$, it is found that the relevant dynamical variables are $j^{\ell}_{\vec{k}A} = j^{\ell}_{\vec{k}C}$ (i.e. the current associated with the concentration variable (3.6c)) and $j^{\ell}_{\vec{k}B} = j^{\ell}_{\vec{k}M}$ (the mass current), whereas in the $k \to \infty$ (independent particle) limit, $\theta(k) \to 0$, as expected.

Once the memory function matrix has been diagonalized at t = 0, it is <u>assumed</u> that it stays diagonal for <u>all</u> times, and a simple exponential or gaussian ansatz is made for the diagonal elements :

$$M'_{\alpha\alpha}(k,t) = \mu'_{\alpha\alpha}(k) \begin{cases} e^{-t/\tau_\alpha(k)} & (6.8a) \\ e^{-t^2/\tau^2_\alpha(k)} & (6.8b) \end{cases}$$

The unknown relaxation times τ_α are chosen to be proportional to the inverse of the characteristic frequencies $\left[\mu'_{\alpha\alpha}(k)\right]^{1/2}$:

$$\tau_\alpha^{-2}(k) = \xi\, \mu'_{\alpha\alpha}(k)$$

where the coefficient ξ is determined so as to lead to the correct free particle limit of $S_{\alpha\alpha}(k,\omega = 0)$ as $k \to \infty$ [13].

The present analysis differs from a similar calculation in ref.[8] in that the Mori continued fraction expansion for the density fluctuation spectra is taken to third rather than to second order ; the present calculation also uses two distinct relaxation times in contrast to the single relaxation time analysis in ref.[8]. The calculated charge fluctuation spectra agree very well with the MD data in the strong coupling regime ($\Gamma > 1$), except as regards the amplitude of the central peak, as shown in fig.2 for $\Gamma = 40$. An analysis based on the high frequency expansion of the correlation functions cannot, a priori, be expected to predict correctly the low frequency behaviour. In fact it is well known that the memory functions contain typically a rapidly decaying "binary collision" part and a more slowly decreasing tail which is essentially determined by mode coupling mechanisms ; such contributions are presently under investigation.

From our analysis we have been able to calculate the spectrum of the concentration fluctuations, $S_{CC}(k,\omega)$. Examples of the spectrum of the associated current correlation function

$$\hat{C}_{cc}(q,\omega) \sim \frac{\omega^2}{q^2} S_{cc}(q,\omega)$$

are shown in fig.3. These spectra exhibit two well-separated peaks for small wavenumbers ($q \lesssim 1$). The high frequency peak is a manifestation of the strong coupling of concentration fluctuations to the plasmon mode, while the low frequency part of the spectrum is related to interdiffusion. With increasing wavenumber the two parts of the spectrum gradually merge into a single broad band.

7. CONCLUSIONS.

Binary ionic mixtures exhibit a very interesting dynamical behaviour which is somewhat intermediate between the time-dependent fluctuations observed in simple metallic alloys (which are dominated by the coupling between concentration and mass fluctuations) and the charge fluctuations of the one component plasma. The phenomenological analysis presented in this lecture yields a satisfactory description of the high frequency phenomena which are dominated by the plasmon mode. A correct description of interdiffusion requires a more detailed mode coupling analysis along the lines of recent work on the electrical conductivity in ionic liquids and plasmas [14].

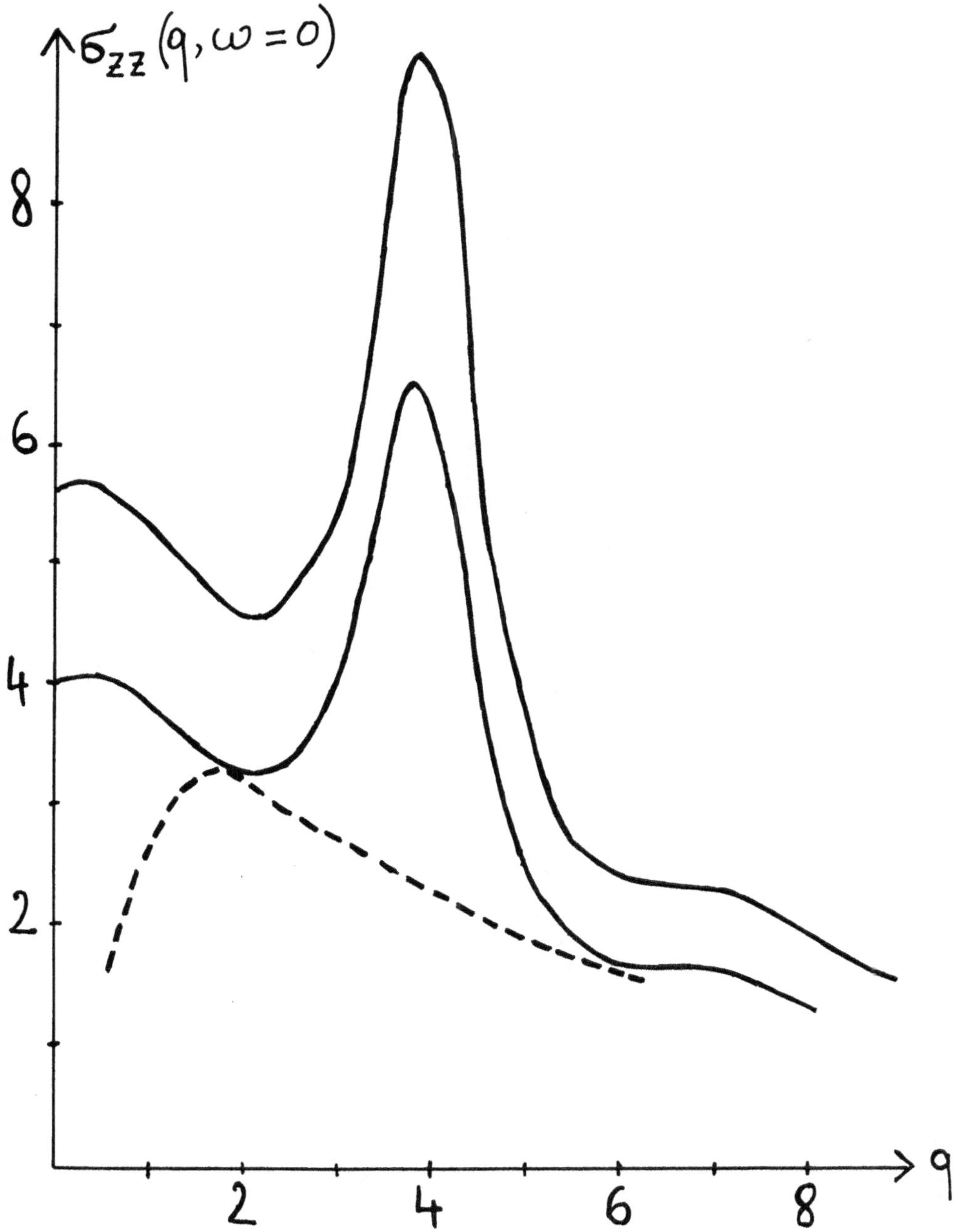

Fig. 2 : $\sigma_{ZZ}(q,\omega=0)$ as a function of q, for a H^+-He^{++} mixture at $\Gamma = 40$, $x_1=x_2= 0.5$. Dashed curve : MD results from ref. [8] ; middle curve : results from exponential memory function analysis ; upper curve : gaussian memory function.

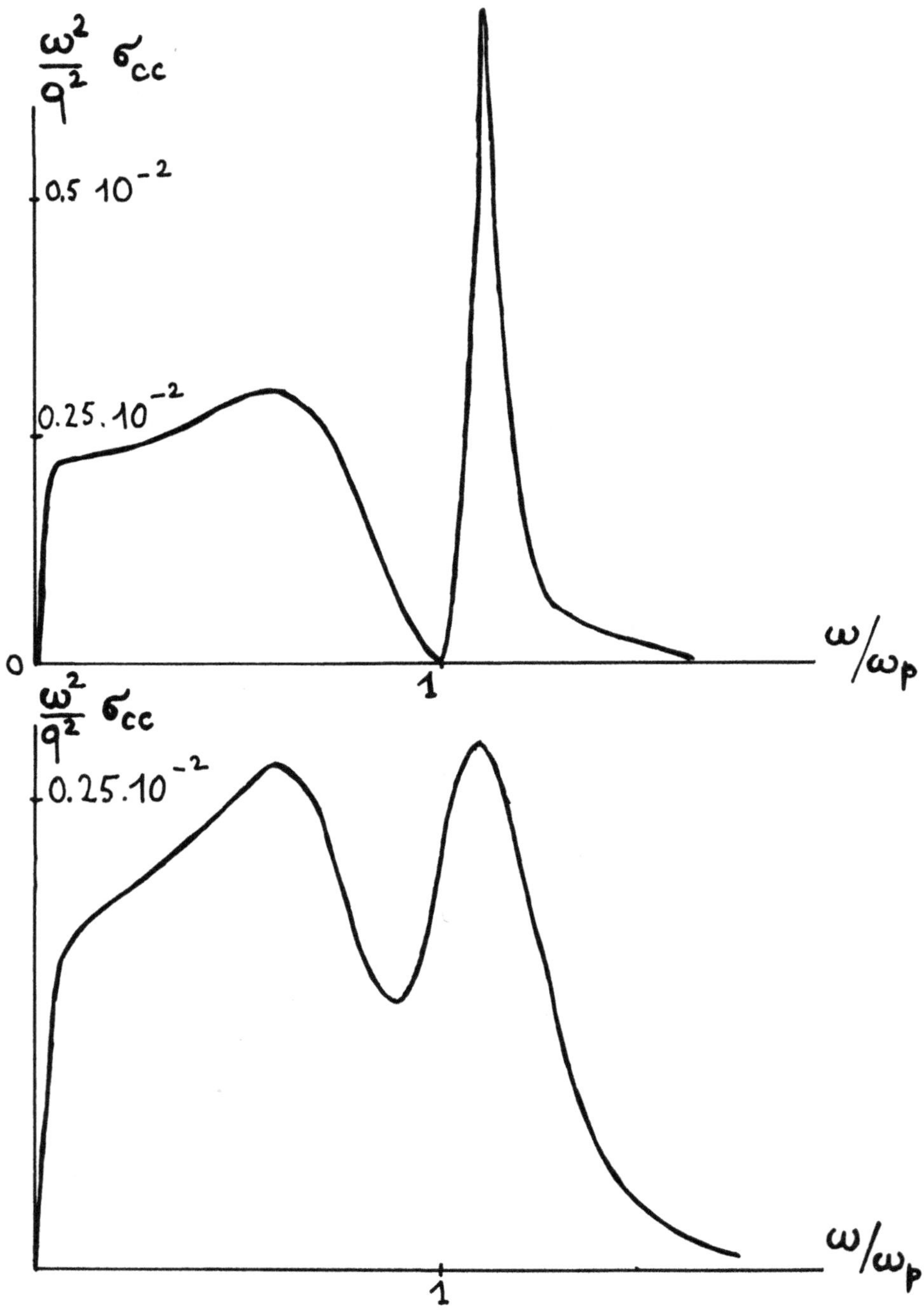

Fig. 3 : $(\omega^2/q^2)\sigma_{CC}(q,\omega)$ as a function of ω/ω_p for a H^+-He^{++} mixture at $\Gamma = 40$, $x_1=x_2= 0.5$. Upper spectrum : q = 0.6; lower spectrum : q = 2. ; the spectra are based on the gaussian memory function analysis.

REFERENCES.

(1) H. Minnoo, C. Deutsch and J.P. Hansen, J. Phys. Lettres 38 (1977) L 191.
(2) D.K. Chaturvedi, G. Senatore and M.P. Tosi, Lett. N. Cimento 30 (1981) 47 ; M.P. Tosi, to be published.
(3) M. Baus and J.P. Hansen, Physics Reports 59 (1980) 1.
(4) D. Bohm and T. Staver, Phys. Rev. 84 (1951) 836.
(5) D. Stevenson, J. de Physique 41 (1980) C2-53.
(6) H.M. Van Horn, J. de Physique 41 (1980) C2-57.
(7) J.P. Hansen, G.M. Torrie and P. Vieillefosse, Phys. Rev. A 16 (1977) 2153.
(8) J.P. Hansen, I.R. Mc Donald and P. Vieillefosse, Phys. Rev. A 20 (1979) 2590.
(9) M. Baus, Phys. Rev. Lett. 40 (1978) 793.
(10) M. Parrinello, M.P. Tosi and N.H. March, J. Phys. C 7 (1974) 2577 ; I. Ebbsjö, P. Schofield, K. Sköld and I. Waller, J. Phys. C 7 (1974) 3891.
(11) J.P. Hansen, I.R. Mc Donald and E.L. Pollock, Phys. Rev. A 11 (1975) 1025.
(12) M.C. Abramo, M. Parrinello and M.P. Tosi, J. Phys. C 7 (1974) 4201.
(13) S.W. Lovesey, J. Phys. C 4 (1971) 3057.
(14) L. Sjögren, J.P. Hansen and E.L. Pollock, Phys. Rev. A 24 (1981) 1544.

Thermodynamics of Ionic Liquids

Konrad G. Weil
Institut für Physikalische Chemie der Technischen Hochschule Darmstadt
Petersenstraße 20, D 6100 Darmstadt

1. Introduction

Within the last two years two review papers have appeared [1,2] which dealt with electrical and structural properties of ionic melts. A third review has been devoted to the density and pressure dependence of the properties of molten salts [3] . The discussions in these reviews of course occured in the framework of macroscopic equilibrium properties of the systems. Therefore, much information on thermodynamics of these liquids is contained in these papers and need not be repeated here.

For this reason this paper is far from being exhaustive. It reviews some recent work dealing with systems which can be located near the border lines between metallic and ionic states and between covalent and ionic states, respectively.

2. Ionic versus Metallic States

2.1 Correlation of Liquid Properties with Molecular Data

The most drastic change in the bond nature of the condensed phases occurs in the series of the alkali metal-gold systems. It appeared to me attractive to look for correlations with bond properties of the corresponding molecules. Using the Knudsen cell mass spectrometry it was possible to measure the bond energy D (T = 0 K) of the molecules LiAu [4] , NaAu [5], RbAu, and CsAu [6].

Usually the ionicity of a chemical bond can be estimated using the electronegativities of the elements. Unfortunately there are no reliable values for the electronegativity of gold. We, therefore, decided to estimate the covalent part of the bond using Pauling's formula

$$D_{cov} = (D_{Me_2} \cdot D_{Au_2})^{1/2} .$$

The energies of dimerization can be taken from a collection by Gingerich [7] . The result of our calculations is presented in Table 1. It can be clearly seen from this table that the condensed phases show ionic properties only in cases when the ionicity of the molecular bond exceeds 50 %. Furthermore it becomes clear that electronegativities alone, even when they are known precisely, are not sufficient for such predictions. Only the relative ionicity is the important quantity.

Table 1

	LiAu	NaAu	RbAu	CsAu
$\frac{D(T=0)}{kJ\cdot mol^{-1}}$	280.8±6.5	212.1±12.6	240.0±3.0	249.0±3.5
$\frac{D_{ion}}{kJ\cdot mol^{-1}}$	127.7	83.7	137.6	155.1
$\frac{D_{ion}}{D(T=0)}$	45%	39%	57%	62%
state of the melt	metallic	metallic	ionic	ionic
state of the solid	metallic	metallic	ionic	ionic

It would be attractive to apply this concept to other systems. Possible candidates are LiTl, NaTl, and LiPb. LiTl and LiPb show anomalies in their electrical conductivities [8]. Further electrochemical measurements on LiPb are discussed later. Solid LiTl and LiPb show the B2 (CsCl) structure, which is a typical metallic structure, while NaTl shows a cubic centered superstructure which may indicate a more pronounced ionic contribution to the solid state properties. The bond energy of the LiPb molecule has been determined by Neubert et al.[9], and those of LiTl and NaTl by Scheuring and Weil [10]. The values are

	LiPb	LiTl	NaTl
$\frac{D(T=0)}{kJ\cdot mol^{-1}}$	74.9±8	75.8±1	68.3±2

Again, as in the case of gold, there are unreliable values for the electronegativities of Pb and Tl. Furthermore, Pauling's simple formula for the covalent contribution to the bond energy cannot be used. The experimental energies of dimerization for Pb and Tl are influenced by multiple bonding. Application of Pauling's formula to the single alkali-thallium bond would extremely overestimate the covalent contribution. Only a very qualitative conclusion can be drawn from the figures given above: the energy of dimerization of Li (101 kJ/mol) substantially exceeds that of Na (71 kJ/mol); while the bond energies of the molecules LiTl and NaTl are virtually the same. This can occur when the ionic contribution to the NaTl bond is greater than that to the LiTl bond.

A more detailed discussion of the polarity of the bond of such intermetallic molecules should be based on quantum chemical considerations. We hope that the availability of very reliable values for bond energies of such molecules will stimulate the theoreticians to perform the necessary calculations.

There is one example where the present state of the art of quantum chemistry is sufficiently developped for such a task: Gollisch [11] has used the density matrix formalism for a relativistic calculation of the ground state properties of CsAu and has obtained agreement between calculated and experimental bond energy which is excellent. He obtains an electron distribution which indicates nearly 90 % ionicity of the bond.

2.2 Thermodynamic Functions

Fig. 1 contains some results on the dependency of thermodynamic functions on the composition in cesium-gold mixtures. The most striking feature of these curves are the steep decrease of cesium activity at very low gold concentrations and the negative value of the partial molar volume of gold in this range of composition. The interpretation given in [13] is straightforward: a gold atom, added to pure or nearly pure cesium, captures an electron from the cesium conduction band. The resulting ion, Au^-, is strongly solvated by cesium ions. The solvation shell is of higher density than the bulk, so that a negative partial molare volume $\overline{V}_{Au}$ results. With increasing gold concentration this effect decreases due to the decreasing number of cesium ions available for solvation.

The decrease of the molar volume of cesium with increasing gold content of the liquid reflects, in my opinion, the continuous change from a metallic state, characterized by a metallic radius of the atom, to an ionic state, characterized by a considerably lower ionic radius of Cs^+. The cesium activity, given in Fig. 1 was calculated from the cesium partial pressures over the liquid. These partial pressures are displayed in Fig. 2. For comparison the corresponding values for the Cs-CsCl system [14] are given. The similarity of both systems cannot be overlooked. From this similarity one interesting question arises: what is the equilibrium cesium pressure over exactly stoichiometric CsAu? Over CsCl it is below the detection limit.Nevertheless the product $p_{Cs} \cdot p_{Cl_2}{}^{1/2}$ is well defined in the presence of stoichiometric CsCl. In the case of CsAu, the product $p_{Cs} \cdot p_{Au}$ depends on the composition of the liquid. The partial pressure of cesium is only uniquely determined, if the gold activity in the system is kept constant.

In [12] also experiments on the cesium activities in the gold rich mixtures are reported. Due to the experimental difficulties the values are not as exact as on the cesium rich side.

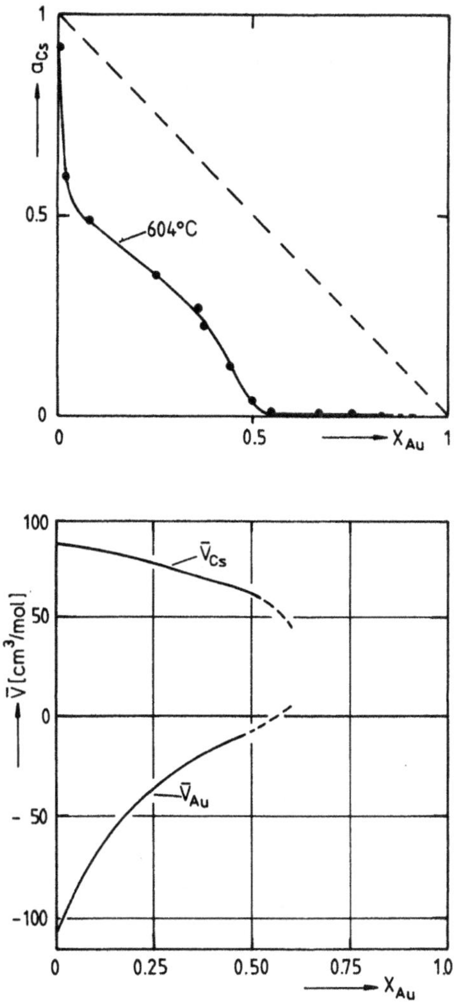

Fig. 1 : a) Cesium activity at 604°C in liquid Cs-Au mixtures [12]
b) Partial molar volumes of cesium and gold in liquid Cs-Au mixtures at 650°C [13]

Nevertheless, together with additional differential thermal analysis results the author can collect a number of arguments in favor of a revised phase diagram Cs-Au. Fig. 3 shows this new diagram. It strongly resembles the Rb-Au phase diagram.

I will not close the section on the CsAu-system without mentioning an interesting result from Martin et al. [15]. These authors measured the density of liquid CsAu mixtures using the method of absorption of γ-radiation, and their results confirm the findings by Kempf and Schmutzler [13]. Using ionic radii for Au^- and Cs^+, determined by neutron diffraction [16], they calculated the molar volume for a dense fluid. In contrary to nearly all alkali halides, this calculated value agrees well with the experimental one. This indicates that vacancies which are typical constituents of alkali halide melts are apparently not present in liquid CsAu. From this point of view it would be interesting to know the volume change at melting for CsAu. This quantity varies between 15 and 28 % for the alkali halides and should be considerably lower for CsAu.

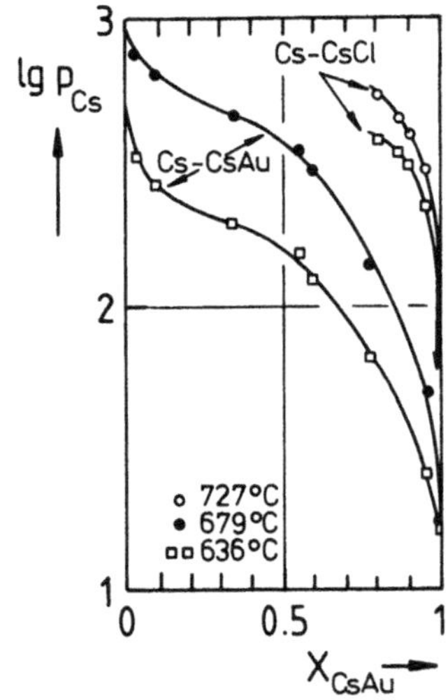

Fig. 2: Cesium partial pressure (Torr) over liquid Cs-CsAu-mixtures [12] and over liquid Cs-CsCl mixtures [14]

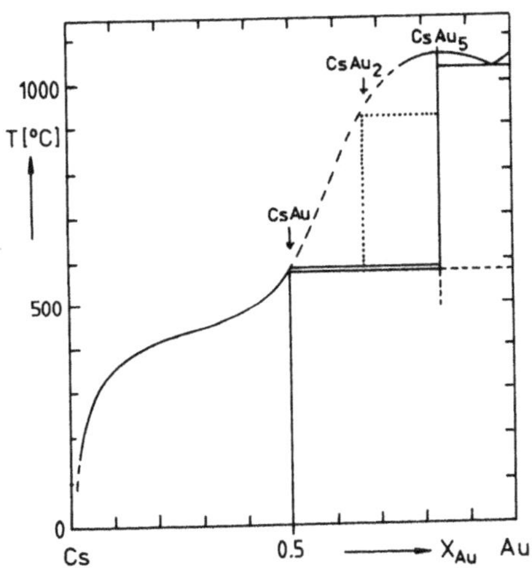

Fig. 3: Revised phase diagram Cs-Au [12]

It is interesting to compare the results obtained with CsAu, which apparently represent an extreme case, with thermodynamic data from a system with less pronounced ionic properties. We choose LiPb as a typical representative for this group. Saboungi, Marr, and Blander [17] conducted EMF-measurements with a Li-electrode in molten Li-Pb-mixtures. From their experiments, they directly obtained the activity coefficients of lithium in the melt. Application of the Gibbs-Duhem-Equation leads to the activity coefficients of lead. Fig. 4 shows their results. There are anomalies in the curves around the composition $x_{Li} = 0.2$, indicating a tendency to form ionic species. This anomaly occurs in the same composition range, where the nimimum in electric conductivity is found [8]. The molar volume also shows a minimum at this concentration [18].

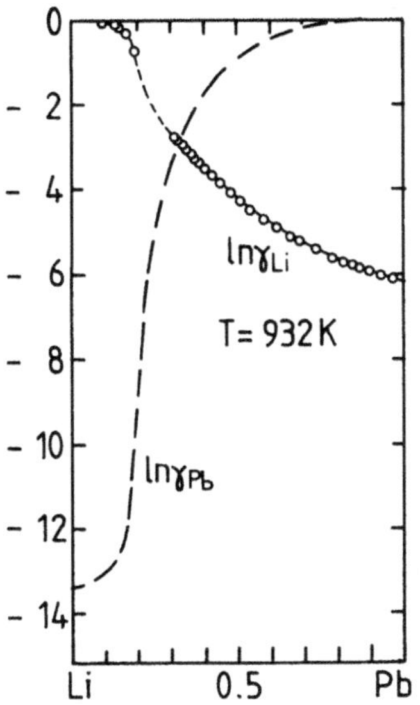

Fig. 4: Activity coefficients of Li and Pb in molten LiPb-mixtures [17]

Charge transfer from one constituent of a mixture to another must lead, of course, to short range order due to coulombic interaction. More information on these systems therefore may be expected from diffraction experiments, which are reviewed in a different session of this conference.

(It seems appropriate to draw the reader's attention to a paper on the excess functions of liquid cesium-cesium halide mixtures which appeared after the presentation of this lecture [24].)

3. Ionic versus Covalent States

In this section I will present some results on properties of molten silver and copper halides. In their review from 1980, Enderby and Neilson [2] reported on neutron diffraction experiments which led to the conclusion that AgCl forms a normal ionic melt. The results which I am going to discuss now give some evidence for structures in these melts which cannot be described by spherical symmetric potentials.

The first clue, that silver chloride , as well as bromide and the corresponding Cu(I) salts are not simple ionic substances can be drawn from the disorder type in the crystalline materials. In contrast to the alkali halides they show Frenkel type disorder. Such an unsymmetric type of disorder cannot be explained on the basis of coulombic intereaction, which is invariant to changes in sign.

The next property of silver chloride which I wish to mention is the volume change at melting. Table 2 shows the exceptional position of this substance.

Table 2

	$\frac{r_+}{pm}$	$\frac{\Delta\overline{V}_m}{\overline{V}_s}$
LiCl	78	0.29
NaCl	98	0.25
AgCl	113	0.08
KCl	133	0.17
RbCl	149	0,14
CsCl	165	0.10

In this table r_+ is the Goldschmidt ionic radius of the cation. $\Delta\overline{V}_m/\overline{V}_s$ is the molar volume change at melting, relative to the molar volume of the solid at the melting temperature. Data come from [19].

The most drastic differences between silver and alkali metals concern the properties of metal solutions in the respective halide melts. Silver and copper are soluble in their molten chlorides and bromides. The electric conductivity of the melt is not changed by the dissolved metal and there is no absorption of light which could be attributed to F-center-like electron states [20].

The solubility is low, and in the temperature range where it can be studied it never exceeds 0.5 mole%. The data follow van't Hoff's law, so that the enthalpy and entropy can be determined from the experiments. The data are quite reliable, for the temperature range from which they were obtained in most cases covered 460°C to 650°C [21]. From the experimental data, using literature values for the sublimation of silver, the enthalpy and the entropy of solvation can be calculated. The standard state for these data therefore is silver gas at 1 atm and the temperature of the experiment.

Striking ist the dependence of these data on the composition of the melt, when mixed halides are used as solvents (Fig. 5). At the present time we cannot interpret these data in terms of molecular models. On the other hand, the specific response of the thermodynamic solvation functions of silver and copper on the composition of the melt seems to me to indicate the presence of molecules or molecular clusters which become part of the solvation shell of the dissolved metal.

This interpretation can be supported by two other independent observations. One comes from my group.We could show that the solvation functions of silver in molten silver chloride react very sensitively to the presence of dopants in small concentrations. Fig. 6 gives a result for $CdCl_2$ as a dopant. Again, enthalpy and entropy of solvation are linearly connected, see Fig. 7.

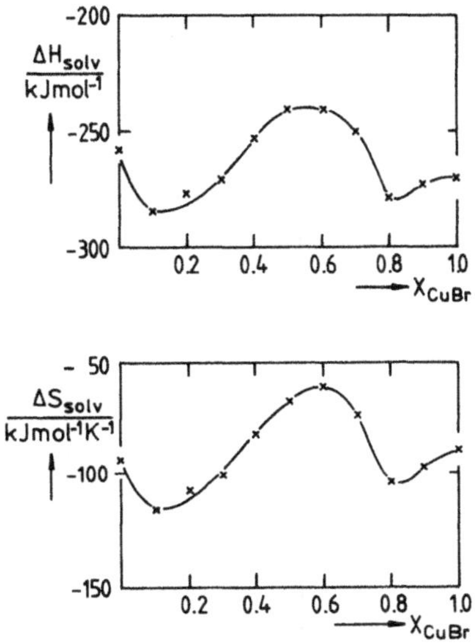

Fig. 5: Enthalpy and entropy of solvation of copper in mixed CuCl-CuBr melts (upper part)

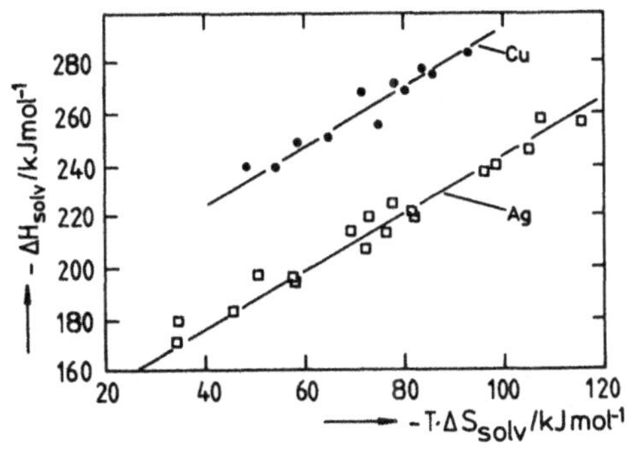

ΔH_{solv} vs. T ΔS_{solv} for silver and copper in their molten halide mixtures[21]

Apparently, this effect is not specific for $CdCl_2$. We tested the following mono and divalent cations at different concentrations: Zn, Cd, Ca, Ba, K, Cs. There seems to be a correlation with the additional charge density in the melt. We define a charge-density-effect by the following expression:

$$C = x_{cat} \cdot (r_{cat}^3 \cdot z_{cat} - r_{Ag^+}^3) \cdot F$$

with x_{cat} = mole fraction of added foreign cation, r_{cat} = its ionic radius,

z_{cat} = its valency, r_{Ag^+} = ionic radius of silver and F = Faraday constant.

We then get the relation displayed in Fig. 8. There seems to be no specifity, and the situation can be summarized as follows: the higher the introduced charge density, the more ionic is the whole liquid.

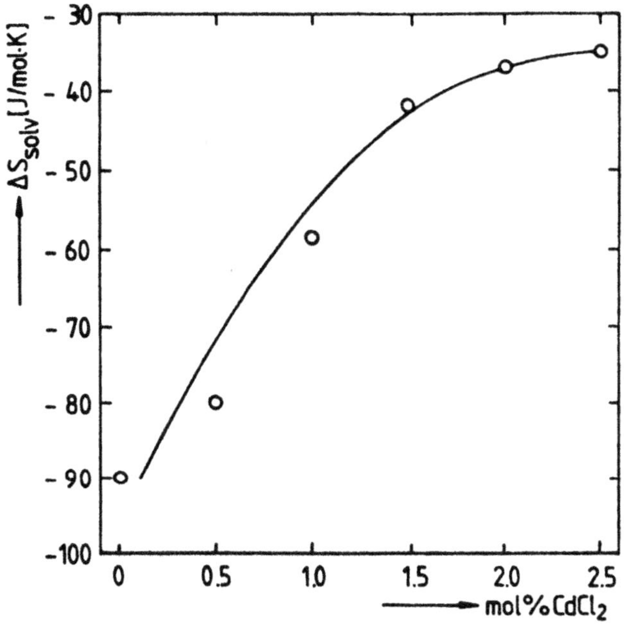

Fig. 6: Entropy of solvation of silver in molten silver chloride vs. concentration of $CdCl_2$ [22]

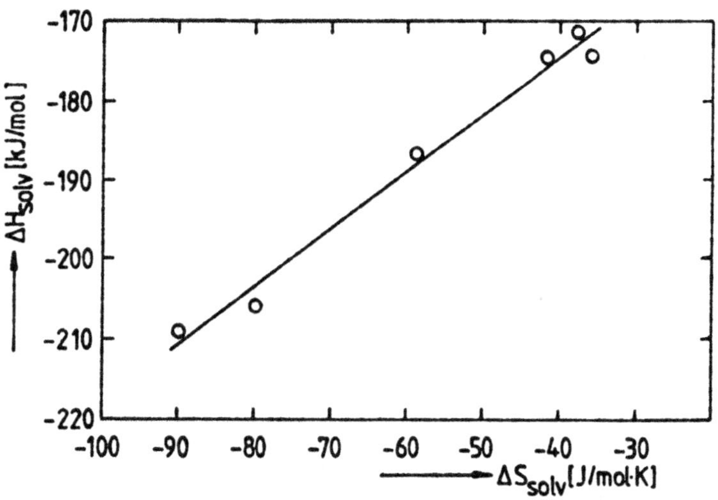

Fig. 7: ΔH_{solv} vs. ΔS_{solv} for silver in molten silver chloride containing different amounts of $CdCl_2$ [22]

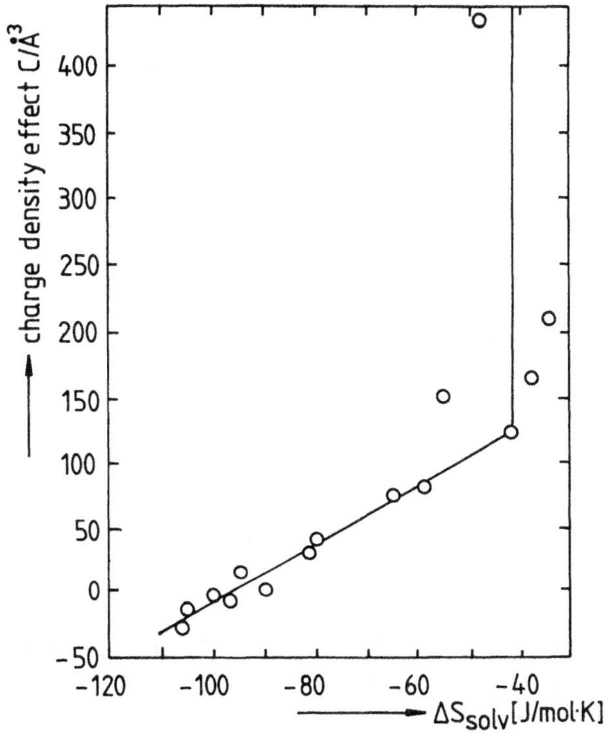

Fig. 8: Entropy of solvation as a function of the charge density effect, defined in the text [22]

As a consequence the silver is not captured to the same extent in solvation shells as in pure silver chloride, it remains more mobile and therefore the entropy of solvation is less negative. There is a saturation effect which may arise when the liquid is fully ionized.

These very tentative conclusions get some support from very recent experiments by Giazitzoglou [23] . From electromotive force measurements he derived excess functions of mixing of AgCl with pure KCl melts and with KCl melts containing 40 mole% LiCl. The excess free energy of mixing with KCl shows no anomaly, but with the Li-containing melt he got the results given in Fig. 9. We see a maximum at $x_{AgCl} = 0.8$. At this composition, the solution contains 8 mole% LiCl and the charge density effect would be above the saturation limit. The charge density effect of potassium, on the other hand, is virtually zero, so that no anomalies in AgCl/KCl mixtures are to be expected.

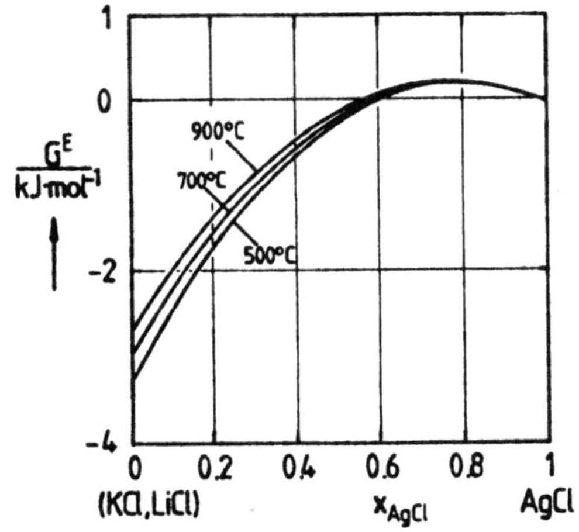

Fig. 9: Excess free energy of mixing of AgCl with KCl containing 40 mole% LiCl [23]

Acknowledgements

I should like to thank Prof. F. Hensel for placing unpublished results at my disposal. Work in my laboratory was sponsored by the Deutsche Forschungsgemeinschaft, the Fonds der Chemischen Industrie and Degussa, Hanau, who supplied us with silver halides. G. Greening was kind enough to help me to overcome some language problems.

Literature references

[1] F. Hensel, Adv. Phys. 28(4), 555 (1979)
[2] J.E. Enderby and G.W. Neilson, Adv.Phys. 29(2), 323 (1980)
[3] K. Tödheide, Angew.Chem. 92, 612 (1980)
[4] A. Neubert and K.F. Zmbov, Trans.Faraday Soc. 70, 2219 (1974)
[5] V. Piacente and K.A. Gingerich, High Temp.Sci. 9, 189 (1977)
[6] B. Busse and K.G. Weil, Ber.Bunsenges.Phys.Chem. 85, 309 (1981)
[7] K.A. Gingerich, Faraday Symposium No. 14, "Diatomic Metals and Metallic Clusters", Paper No. 5, Manchester 1980
[8] V.T. Nguyen and J.E. Enderby, Philos.Mag. 35, 1013 (1977)
[9] A. Neubert, H.R. Ihle, and K.A. Gingerich, J.Chem.Phys. 73, 1406 (1980)
[10] T. Scheuring and K.G. Weil, Adv.Mass.Spectrom., to appear 1983
[11] H. Gollisch, J.Phys. B, in press
[12] A. Kempf, Dissertation, Marburg 1981
[13] A. Kempf and R.W. Schmutzler, Ber.Bunsenges.Phys.Chem. 84, 5 (1980)
[14] R.L. Mc Graw and N.L. Nachtrieb, quoted in [12]
[15] W. Martin, N.M. Keita, P.Lamparter, and S. Steeb, J.Phys.Chem.Liq. 10,84 (1980)
[16] W. Martin, W. Freyland, P. Lamparter, and S. Steeb, J.Phys.Chem.Liq. 10, 61(1980)
[17] M.-L. Saboungi, J. Marr, and M. Blander, J.Chem.Phys. 68, 1375 (1976)
[18] H. Ruppersberg and W. Speicher, Z.Naturforschg. A 31, 47 (1976)
[19] Molten Salts Handbook, Academic Press, London 1967
[20] A.F. Schwind, R. Vötsch, and K.G. Weil, Z.Naturforsch. 36a, 354 (1981)
[21] E. Schuster, A.F. Schwind, R. Vötsch, and K.G.Weil, Z.Naturforsch. 34a,1203(1979)
[22] T. Lieser and K.G. Weil, to be published
[23] Z. Giazitzoglou, Ber.Bunsenges.Phys.Chem., in press
[24] H. Yokokawa and O.J. Kleppa, J.chem.Phys. 76 (11), 5574 (1982)

THE STRUCTURE OF IONIC MELTS

J.E. Enderby
H.H. Wills Physics Laboratory
University of Bristol

Bristol BS8 1TL, U.K.

1. Introduction

The structural properties of a molten salt or a liquid semiconductor containing cations (+) and anions (-) are characterised by three partial structure factors $S_{++}(k)$, $S_{--}(k)$ and $S_{+-}(k)$ and three pair correlation functions $g_{++}(r)$, $g_{--}(r)$ and $g_{+-}(r)$. The pair correlation functions and the partial structure factors are related through expressions of the form

$$g_{\alpha\beta}(r) = 1 + \frac{1}{2\pi^2 \rho r} \int S_{\alpha\beta}(k) k \sin kr\, dk$$

where ρ is the total number density. The partial structure factors contribute to the total structure factor $F(k)$, a quantity accessible from a single X-ray or neutron diffraction experiment, according to:

$$F(k) = c_+^2 f_+^2 (S_{++}(k)-1) + c_-^2 f_-^2 (S_{--}(k)-1) + 2c_+c_-f_+f_-(S_{+-}(k)-1)$$

Here $c_{+(-)}$ is atomic fraction of the cation (anion), and $f_{+(-)}$ is the scattering amplitude of the cation (anion) appropriate to the radiation used. It is widely accepted that, where possible, S_{++}, S_{--} and S_{+-} should be extracted from the experimental data and this can in principle be accomplished by changing the $f_{+(-)}$ values in a systematic way.

2. Molten NaCl

In 1975 Edwards et al (1975) published the results of the first full three-pattern neutron experiment on molten NaCl. In this experiment the isotopic state of the chloride ion was changed and in this way three distinct scattering patterns were obtained. From their results Edwards et al were able to identify some general properties of $S_{\alpha\beta}(k)$ and $g_{\alpha\beta}(r)$ for 1-1 ionic melts which have since been subsequently confirmed for many comparable systems. These include:

(a) a 'coulomb' peak in $S_{++}(k)$ and $S_{--}(k)$ at $k \sim 2Å^{-1}$ and a corresponding 'coulomb' dip in $S_{+-}(k)$ at the same value of k

(b) an overall similarity between $g_{++}(r)$ and $g_{--}(r)$

(c) a tendency for $g_{++}(r)$, $g_{+-}(r)$ and $g_{--}(r)$ to be phased so that for $r \gg 4Å$ the cancellation of charge is complete

(d) a small but finite amount of penetration by the like species into the first coordination shell.

Some specific properties of molten NaCl found by Biggin and Enderby (1982) in their re-evaluation of the data with more reliable values for f_- were:

(a) a reduction in the Na-Cl interionic distance r_{NaCl} (as determined by the position of the first peak in $g_{+-}(r)$) from 2.95Å (solid) to 2.78Å (liquid)

(b) a coordination number for cations around anions, n_{+-}, of 5.3±0.4 derived by integrating $g_{+-}(r)r^2$ from zero to the first minimum in $g_{+-}(r)$

(c) no significant differences between $g_{++}(r)$ and $g_{--}(r)$, within the experimental error.

3. Other Molten Salts

A summary of experimental data for molten $ZnCl_2$, $CaCl_2$ and $BaCl_2$ is shown in table 1 below.

Table 1

Salt (ionic radius) (Å)	Cation-Cation		Cl-Cl		Cation-Cl	
	$\bar{r}_{++}$ (Å)	$\bar{n}_{++}$	$\bar{r}_{--}$ (Å)	$\bar{n}_{--}$	$\bar{r}_{+-}$ (Å)	$\bar{n}_{+-}$
$ZnCl_2$ (0.74)	3.8±0.1	4.7±0.8	3.71±0.02	8.6±0.02	2.29±0.02	4.3±0.3
$CaCl_2$ (0.99)	3.6±0.1	4.2±0.5	3.73±0.03	7.8±0.3	2.78±0.03	5.4±0.3
$BaCl_2$ (1.35)	4.9±0.1	14 ± 2	3.86±0.02	7 ± 1	3.10±0.02	7.7±0.2

References. $ZnCl_2$: Biggin & Enderby (1981a); $CaCl_2$: Biggin & Enderby (1981b); $BaCl_2$: Edwards et al (1978).

Biggin and Enderby (1981a,b) have pointed out that the gross structural properties for most salts can be understood in terms of relative sizes of the cation and anion. In particular, they have shown that once the radius ratio r_-/r_+ exceeds 2.4 there is little or no penetration of the first coordination shell by like ions. Even molten $ZnCl_2$, which has somewhat unusual transport properties, can be structurally well represented by a simple ionic model. This general rule applies unless there are specific chemical effects which modify the bonding in a crucial way as, for example, in copper or nickel-based systems like, for example, liquid CuCl.

4. Liquid Semiconductors

Consider a liquid alloy of the form M_aA_m where M is the more electropositive element of valence m and A, the more electronegative element, has valence a. If the elements M and A are metallic in the pure state but have very different electronegativities, the liquid alloy is usually a 'liquid semiconductor' i.e. it is characterised by a non-metallic value of the conductivity, whose temperature coefficient is positive. Well-known examples of such behaviour are to be found in liquid Mg_3Bi_2, Li_4Pb, Li_3Bi and CsAu (Cutler 1977). For these systems a model based on ionic bonding seems capable of explaining most of the observations; this model, first explicitly proposed by Enderby (1973) has been more fully justified by Roth (1976) and seems to have gained general acceptance (Robertson 1979).

The situation for Te-based alloys was, until recently, less clear. For liquid $NiTe_2$, NiTe and CuTe evidence has been given by Nguyen, Gay, Enderby, Newport and Howe, (1982) and Newport, Howe and Enderby (1982) that substantial electron transfer takes place and, in the stoichiometric alloy, the tellurium is in the form of Te^{2-} ions. The comparatively high conductivity of these liquids is postulated to arise from the d-bands associated with the d^8 (Ni^{2+}) and d^9 (Cu^{2+}) cations, such bands being on the delocalised side of the Mott transition (Wilson 1972). A summary of the relevant structural data is shown in table 2.

Table 2

	Average Coordination Numbers			Distances (Å)		
	M-M	Te about M	Te-Te	M-M	Te-M	Te-Te
NiTe	2.9±0.4	4.5±0.5	10±3	2.54±0.03	2.56±0.03	3.56±0.04
$NiTe_2$	1.6±0.4	4.6±0.3	11±3	2.81±0.03	2.56±0.03	3.55±0.04
CuTe	3.6±0.4	2.9±0.3	10±3	2.82±0.03	2.55±0.03	3.66±0.05

We can see that a nickel ion is, on average, surrounded by ∿ 4 Te^{2+} ions at 2.56Å and ∿ 3 Ni^{2+} ions at 2.54Å in liquid NiTe. This local order provides the essential clue to the electrical behaviour reported in the paper by Newport et al (1982). The Te^{2-} ions form a close-packed structure in much the same way as Cl^- does in a range of molten chlorides. However, as stated earlier, for a radius ratio > 2.4 no cations can penetrate the first coordination shell for a simple molten salt. By contrast, the cations in these alloys clearly do penetrate the first coordination shell and are able to form a three-fold coordinated structure with a Ni-Ni or Cu-Cu spacing sufficiently short to allow significant d-band overlap. It is our view that the d-character of nickel and copper plays a crucial role in determining the bonding and electrical characteristics of these liquid tellurides. Experiments are in progress on liquid MnTe which, according to this picture, should have a similar structure to NiTe but, on account of the Mott transition, have a much smaller conductivity.

One interesting observation concerns liquid Cu_xTe_{1-x}; for x > 0.5, the conductivity decreases rapidly and becomes a minimum at Cu_2Te. By contrast, liquid Ni-Te continues its gradual progression from semi-metallic (liquid Te) to metallic (liquid Ni) behaviour. The presence of monovalent Cu in the d^{10} configuration implied by the stoichiometry of Cu_2Te will lead to low conductivities; no such behaviour will occur for Ni-Te and the model proposed therefore gives a natural explanation of the contrasting behaviour for x > 0.5 of the two alloy systems.

The total structure factor of liquid Tl_2Te has recently been measured by Gay, Copestake and Enderby (1982) and compared with the predictions in k-space of the mean spherical model of an ionic liquid consisting of Tl^+ and Te^{2-} ions. Good overall agreement was achieved. The total radial distribution function in real space was also investigated and it was concluded that models of the melt based on discrete molecules should be abandoned and that an ionic model is a most useful starting approximation. However, it is sometimes argued that the electronegativity difference $\Delta\chi$ is too small for an ionic model to be valid for metal-tellurium alloys of this type. For Tl_2Te $\Delta\chi$ is 0.6 (Cutler 1977). The question of ionicity versus covalency can, as is now widely recognised, be answered only in terms of charge density maps. Complete ionicity in the sense that all the charge is located on the cations and the anions does indeed involve large values of $\Delta\chi$ (in NaCl, for example $\Delta\chi = 2.2$). However, the detailed charge density maps calculated by Schutler et al (1975) for solid PbTe and PbSe ($\Delta\chi$ = 0.5 and 0.8 respectively) show that to an excellent approximation most of the charge is indeed located on Pb^{2+} cations and Te^{2-} or Se^{2-} anions. It is unlikely that melting PbTe or PbSe would actually increase the covalent character of this compound so that liquid PbTe and PbSe will almost certainly be characterised by substantial electron transfer. We therefore conclude that a comparatively small value of $\Delta\chi$ does not, in itself, preclude partial ionicity of the sort we propose for liquid Tl_2Te. The effective valence of the cation, however, is important and the model advanced in our work applies only to monovalent (Cu^+, Tl^+, Ag^+) and divalent (Cu^{2+}, Ni^{2+}, Sn^{2+} and Pb^{2+})cations . Once the 8-N rule is satisfied by both components, as for example in liquid or glassy $GeSe_2$, quite different physics is involved. Liquid In-Te, Ga-Te and Cs-Sb represent interesting intermediate cases and would benefit from the detailed structural studies of the type reported here.

Acknowledgements

The structural studies were carried out at AERE Harwell and we are grateful for the continued support of the SERC. I wish to thank John Wilson, Susan Biggin and Bob Evans for many helpful conversations.

References

Biggin, S. and Enderby, J.E. 1981a, J. Phys. C. Solid State Phys. 14, 3129
Biggin, S. and Enderby, J.E. 1981b, J. Phys. C. Solid State Phys. 14, 3577
Biggin, S. and Enderby, J.E. 1982, J. Phys. C. Solid State Phys. 15, L355
Cutler, M. 1977, Liquid Semiconductors (Academic Press : New York)
Enderby, J.E. 1973, Band Structure Spectroscopy of Metals and Alloys (Academic Press: London) p. 609
Edwards, F.G., Enderby, J.E. and Page, D.I. 1975, J. Phys. C. Solid State Phys. 8, 3483
Edwards, F.G., Howe, R.A., Enderby, J.E. and Page, D.I. 1978, J. Phys. C. Solid State Phys. 11, 1053
Gay, M., Copestake, A.P. and Enderby, J.E. 1982, J. Phys. C. Solid State Phys, in press
Newport, R.J., Howe, R.A. and Enderby, J.E. 1982, J. Phys. C. Solid State Phys, in press
Nguyen, V.T., Gay, M., Enderby, J.E., Newport, R.J. and Howe, R.A. 1982, J. Phys. C. Solid State Phys, in press
Robertson, J. 1979, Phil. Mag. 39, 479
Roth, L.M. 1976 Liquid Metals (Institute of Physics : London) p. 206
Schutler, M., Martinez, G. and Cohen, M.L. 1975, Phys. Rev. 11, 3808
Wilson, J. 1972 Adv. Phys. 21, 143

STRUCTURAL PROPERTIES OF MOLTEN-SALT SYSTEMS

M.P. Tosi

Theoretical Chemistry Department
University of Oxford

1. Introduction

The past decade has seen great progress in the microscopic understanding of the structure of molten alkali halides. This has been mainly brought about by the use of neutron diffraction techniques to resolve the partial structure factors for the component species from the total diffraction pattern of the liquid [1-4]. Computer simulation results for model systems, when contrasted with these data, show that the interionic forces in these liquids near freezing are very similar in character to those operating in the solid phase [5]. A rough description of the main structural features can thus be given at the most elementary level in terms of simple physical parameters, the ionic valences and the ionic radii [6]. While the determination of partial structure factors by neutron diffraction has so far focussed on the family of the chlorides, other systems which may usefully be investigated by alternative experimental techniques have already been covered by computer simulation work, on the basic assumption that the interionic forces are in essence known from solid-phase data [7]. Computer simulation methods have also been very useful in exploring the dynamical and transport properties of these systems [5,8] - an area where experimental methods such as neutron inelastic scattering [9] and optical reflectivity measurements [10] have so far been applied rather sporadically.

An extensive review of recent progress in this field is already available in the literature [11]. The present article instead, after a summary on the main structural features of the pure molten alkali halides in section 2, aims at giving a report on progress which is being promoted in related areas by the above advances. The topics to be taken up in succession are the relation between the structure of a molten salt near freezing and its freezing transition in section 3, the nature of electronic traps and the optical absorption spectrum in dilute solutions of alkali metals in molten salts in section 4, an overall preliminary view of the structure of metal-molten salt solutions in section 5, and structural properties of ionic impurities and molten salt mixtures in section 6. Most of these areas are of current experimental interest - the discussion will focus on interpretations in terms of simple models.

2. Structure of molten alkali halides

The main structural characteristic of a molten alkali halide near freezing is the strong short-range order in the charge density correlations. The liquid is structured around a given ion in successive shells of ions with alternating charges, which are seen from the central ion as progressively broadening and interpenetrating with increasing distance. In wavenumber space this state of order is best brought up directly from the measured partial structure factors by building from them linear combinations which give the autocorrelation $S_{QQ}(k)$ of the charge density fluctuations and the autocorrelation $S_{NN}(k)$ of the total particle density fluctuations. In a liquid like RbCl [4], where the differences in physical parameters between the two ionic species are rather minor except for the opposite valences, the cross-correlations $S_{NQ}(k)$ are correspondingly minor. Such a system can then be viewed as composed of two interpenetrating and approximately independent "liquids".

The first "liquid", whose structure is described by $S_{QQ}(k)$, is composed by the charge density fluctuations. $S_{QQ}(k)$ is qualitatively similar to the structure factor of a dense, strongly coupled classical plasma. It has a strong sharp peak at a wavenumber (k_Q, say) corresponding to the first star of reciprocal lattice vectors of the fcc Bravais lattice appropriate to the rocksalt structure of the solid phase, and drops rapidly to very small values towards the small-angle scattering region where according to the theory it should vanish as k^2/k_{DH}^2 by charge neutrality. This sharp structural feature is hardly apparent in the total diffraction pattern because the corresponding form factor involves the difference in form factors of the two ionic species (the odd indexed Bragg reflections of the rocksalt structure are known to crystallographers as "forbidden" reflections).

The second "liquid", whose structure is described by $S_{NN}(k)$, is composed by the total particle density fluctuations. $S_{NN}(k)$ has a shallow broad peak at a wavenumber $k_N \cong 1.4k_Q$, which would remind one of a monatomic fluid like argon in a real-gas thermodynamic state except for a shoulder on its high-k side. The shoulder lies at a wavenumber approximately equal to $2k_Q$, suggesting a modulation induced by the strong charge ordering. Furthermore, and unlike the structure factor of a gas, $S_{NN}(k)$ drops gently to small values towards the small-angle scattering region. According to the theory $S_{NN}(k)$ should be proportional for $k \to 0$ to the isothermal compressibility, which has of course a liquid-like value in the dense molten salt.

For a system like NaCl [2], where the size difference between the two ionic species is already quite appreciable, the above qualitative picture remains useful but must be supplemented with the observation that the degree of delocalization of the smaller-sized ion (the Na^+ ion in this case) is appreciable higher than that of its partner.

An important consequence of the strong charge ordering in the molten salt is that the average value of the Madelung potential seen by an ion is not strongly decreased from the value appropriate for an ion at a lattice site in the crystal. This point can be gauged most simply by comparing the cohesive energies of the liquid and solid phase, since this quantity is within roughly 10% determined in an ionic model by the Coulomb interactions between the ionic charges. The experimental internal energy of molten NaCl at 1073K, for instance, is 711×10^{3} J mol^{-1} against a value of 764×10^{3} J mol^{-1} for the solid phase at 298K. The structural data also show that the drop in coordination number on melting is rather small.

On the other hand the low compressibility of the molten salt reflects principally the short-range overlap repulsions between the ions. This fact was exploited in ref. [6] to deduce a set of ionic radii in a charged-hard-spheres model. By combining liquid-state compressibility data with information on ionic radii from solid-phase cohesive data [12], it was possible to show that an ionic radius with a weak temperature dependence is a characteristic parameter of each ion, independently of its partner. The whole set of size parameters for the family of molten alkali halides is thus reduced to just one disposable parameter, in essence the scale of length measuring the ionic sizes, which depends on the thermodynamic state once the ionic interactions are simulated by a charged-hard-spheres model. This approach also yields a good semiquantitative description of charge ordering in the molten salt.

We proceed from the above essential facts for the pure molten alkali halides to a discussion of more complex physical problems.

3. Freezing of molten salts

An empirical connection between the freezing transition and the height attained on cooling by the main peak of the structure factor has been known for quite some time for monatomic liquids [13] and has recently found a statistical-mechanical justification in the work of Ramakrishnan and Yussouff [14]. The Fourier components $\rho(\underset{\sim}{G})$ of the periodic atomic density in the solid phase, $\underset{\sim}{G}$ being the reciprocal lattice vectors, can be taken as the order parameters for the transition. A large value for S(k) on the other hand implies, by the classical fluctuation-dissipation theorem, a small value for the free energy spent in modulating the homogeneous liquid density with density waves of corresponding wavevectors $\underset{\sim}{k}$. A first-order transition from the homogeneous to a modulated phase can be obtained in a simple liquid by balancing the loss of free energy incurred in the modulation process against a free energy gain arising from volume contraction.

In extending these considerations to an ionic crystal with the rocksalt structure it is useful to take the Fourier components $\rho_Q(\underset{\sim}{G})$ and $\rho_N(\underset{\sim}{G})$ of the charge density and of the total particle density as its order parameters [15]. Then $\rho_N(\underset{\sim}{r})$ describes the sc lattice to which the rocksalt structure reduces if the differences between the two ionic species are ignored. $\rho_Q(\underset{\sim}{r})$ instead reflects the alternation of the ionic charges in the crystal structure and characterizes the ionic crystal as an fcc lattice with a basis. The corresponding autocorrelation functions in the liquid phase are the structure factors $S_{QQ}(k)$ and $S_{NN}(k)$ discussed in the preceding section and one expects that the height of the sharp peak in $S_{QQ}(k)$ will be an important parameter in the freezing of the alkali halide. This is illustrated in table 1, from ref. [15], on the basis of the neutron diffraction data that are available [2-4] for the molten chlorides. These measurements refer to different temperatures T relative to the observed freezing point T_m, as shown in the table, but a rough extrapolation to the freezing point

Table 1. Height of principal peak in S_{QQ} in molten alkali chlorides

	NaCl	KCl	RbCl	CsCl
S_{QQ}^{max}	3.2	4.4	3.6	3.1
$(T-T_m)/T_m$	0.070	0.023	0.035	0.059
$e^2/(ak_BT_m)$	71	65	66	65

leads to the estimate

$$S_{QQ}^{max} \sim 5$$

at freezing. One must caution however that there is a variation of the ratio of ionic radii across the systems reported in the table and, although its effect is at present not known, one knows that wide variations can lead to different crystal structures. The table also shows that an alternative criterion for freezing in these systems is

$$e^2/(ak_BT_m) \cong 70,$$

a being the mean interparticle distance. A related criterion was pointed out by Reiss et al [16].

The Ramakrishnan - Yussouff theory of freezing for a monatomic system is easily extended to the case of a molten salt, in terms of the order parameters $\rho_Q(\underset{\sim}{G})$ and $\rho_N(\underset{\sim}{G})$ [17]. At the most elementary level the freezing process can be described as a first-order phase transition in which the molten salt is modulated with charge density waves at wavevectors $\underset{\sim}{k}_Q$ (corresponding to the first star of reciprocal lattice vectors), the necessary free energy being supplied from the rather large volume contraction on freezing. The corresponding structural parameters which govern the transition are $S_{QQ}(k_Q)$ and $S_{NN}(0)$. Preliminary calculations on RbCl by M. Rovere indicate that, while this suffices to obtain a transition to a periodic phase with a volume change of the correct order of magnitude, additional components of the set of order parameters, and in particular those corresponding to the main peak of $S_{NN}(k)$, will have to be included in an attempt to make full quantitative contact with experiment.

It is also interesting to note that a similar analysis for $BaCl_2$, which is an anionic-type superionic conductor in the fluorite-type solid phase at high temperatures, suggests a rather different freezing mechanism. The liquid structure data for this system [18] seem compatible with a mechanism in which the cation sublattice freezes with the volume change associated with the first-order phase transition and in so doing induces an average change and a modulation of the number density of the anions [19]. In particular from these data and the measured volume change one estimates

$$\rho_{Cl}(\mathcal{G}_1) \cong 0.4\, \rho_{Ba}(\mathcal{G}_1)$$

for the relative magnitude of the Fourier components of the singlet densities of anions and cations at the first reciprocal lattice vector.

4. F-centre absorption in molten salts

Broad bands of optical absorption at the low frequency end of the visible spectrum are observed in very dilute solutions of alkali metals in their molten halides [20]. The spectroscopic evidence strongly suggests a bound electronic state in the liquid akin to the F-centre in the crystal, in which the electron is bound in a cavity inside the structure where it replaces a halogen ion [21]. The absorption peak frequency and the absorption band width in the liquid join smoothly with the F-band peak and width in the crystal when plotted against the two-thirds power of the density and against the heat content of the system, respectively, in spite of the discontinuous change in these quantities across the melting transition [22].

A theoretical treatment of this type of electron trap in the molten salt hinges on a selfconsistent determination of the electronic wave function and of the surrounding liquid structure [23]. Assuming that the electronic distribution is centred inside a spherical bubble whose diameter is ultimately to be determined by minimizing the free energy, the electronic charge enclosed in the bubble determines the surrounding liquid structure which in turn determines the Madelung potential binding the electron in the bubble. This potential for a halogen ion in the molten salt is quite a healthy one, as was remarked in section 2 above. When the halogen ion is replaced by an electron, the kinetic pressure of the latter induces local readjustments of the structure, in the form of an outward expansion of the bubble, as well as a spill-out of the electronic distribution. This could in principle lead to a dissolution of the local structure and to the disappearance of the bound state. The Madelung potential in the liquid is sufficiently strong, however, that only minor adjustments are found for the 1s electronic ground state : for KCl near freezing one estimates that the bubble diameter expands by roughly 10% and that roughly 80% of the electronic charge is still contained inside the bubble diameter.

The corresponding optical absorption band is then calculated on the assumption of a Franck-Condon transition from the 1s to the 2p state, by allowing for electronic relaxation but not for ionic relaxation during the transition. The band width arises in this model from spherically symmetrical equilibrium fluctuations of the ionic liquid around the bubble in the electronic ground state. One finds for KCl an absorption peak frequency of roughly 1.5 eV (against a measured value of 1.33 eV [24]), whereas the band width is underestimated by a factor of two. The same model allows one to establish an empirical connection between the optical absorption spectrum and the ionic radii in the various molten alkali halides, which is in accord with the available observations.

The discrepancy between the calculated and the observed band width could of course be indicative of an important role of non-spherically symmetrical fluctuations in the electronic ground state and possibly of F-centre aggregation phenomena already occurring at rather high dilution. Calculations in progress by E. Roman however yield a band width which is appreciably closer to experiment by considering just the following two effects:

(i) The explicit dependence of the long-range potential tail outside the bubble on the electronic charge inside the bubble tends to increase the probability of spherically-symmetric fluctuations relative to the earlier calculations;

(ii) transitions to higher excited states contribute to the optical absorption and in particular the 1s → 3p transition is estimated to occur in KCl at an energy higher than that of the 1s → 2p transition by roughly 0.4 eV, with a relative probability of the order of 10%. This preliminary result appears to be in qualitative accord with experimental data on CsCl by Nachtrieb [25], although more work is needed to make more precise contact between theory and experiment.

One should also point out that the estimated fraction of electronic charge inside the unrelaxed ground-state bubble is only 50% for the 2p state and a few percent for the 3p state. As the ionic system is allowed to relax after the assumed Franck-Condon transition, there is clearly the possibility that the electron will become unbound.

5. Structure of metal-molten salt solutions

Whereas the above discussion refers to the ideal case of infinite dilution, alkali metals and molten salts form true solutions over the whole range of composition in the high-temperature liquid phase (for early reviews see Bredig [26] and Corbett [27]). An interesting physical property of these systems obviously is their electric conductivity, which changes with composition from that of an ionic liquid to that of a liquid metal [28]. In the salt-rich region a thermally activated electronic contribution to the conductance can be approximately isolated from the data: various interpretations [29,30] imply the existence of localized electronic states in these solutions, of which the F-centre bound state discussed above is a specific example at very high metal dilution. In the metal-rich region a strong enhancement of the resistivity over that of the pure liquid metal is observed: this has been interpreted at high salt dilution [31] as due to volume expansion and to local restructuring of the liquid metal by the halogen-ion impurities. The behaviour of the conductivity in the intermediate region of concentration, where a mixibility gap appears in many of these systems below a critical temperature, should be particularly interesting but has not so far been studied in any detail. By combining measured values of the conductivity with the Mott criteria, one expects that the transition from non-metallic to metallic behaviour should occur in this region of concentration.

An extensive study of the structure of these solutions by neutron diffraction techniques has been initiated by the Lyon-Grénoble group [32,33]. In the salt-rich region (for the case of Rb_x-$(RbBr)_{1-x}$ with $x = 0.15$) the total diffraction pattern is very similar to that of the pure molten salt, except for some additional scattering in the small-angle region which may be due to the presence of small metallic aggregates. Very pronounced structural changes relative to the pure liquid metal are instead observed in the metal-rich region, for the case K_x-$(KCl)_{1-x}$ with $0.75 \leqq x \leqq 0.95$. The main peak region of the total diffraction pattern up to rather large values of x ($x \cong 0.9$) can be interpreted in this system [34] in terms of the same basic ionic model that applies to the pure molten salt. At higher concentrations of metal, instead, the main peak region of the total diffraction pattern can be viewed as reflecting an expanded metal strongly disordered by the halogen ions, in full accord with the above mentioned interpretation [31] of the resistivity data. The transition between these two structural situations, which correspond to a cooperative structuring effect by the halogen-ion component on the one hand and to each halogen ion acting as an impurity to construct its coordination shell on the other, appears to occur in a rather narrow range of composition, lying around $x \cong 0.9$ in K-KCl and $x \cong 0.8$ in Rb-RbBr.

The neutron diffraction data also show a strong enhancement of the small-angle scattering in the metal-rich region. Unpublished calculations by G. Senatore show that simple electron-screening theory of the electron-ion mixture representing these solutions can account for this extra scattering as far as the total diffraction pattern is concerned. However, this simple approach does not reproduce the qualitative behaviour of the partial structure factors, which have been experimentally resolved [33], although with appreciable statistical uncertainties, from the total diffraction pattern in K_x-$(KCl)_{1-x}$ at $x = 0.8$. In particular the data indicate a pronounced peak in the K-K partial structure factor in this region of wavenumber. This feature if confirmed would seem to suggest cooperative ordering by negative ions and electrons rather than simple electronic screening of ionic density fluctuations.

It should be clear from the above discussion that the experimental picture of these systems is by no means complete and that much remains to be done to establish contact between theory and experiment. Further intensive work may be expected in the coming years, in view of the high current interest in non-conventional ionic liquids [35] of which the metal-molten salt solutions are a relatively simple example from the theoretician's point of view. It may also be appropriate to mention at this point that interest in these electron-ion fluids has stimulated an analysis of the structure of the liquid alkali metals from an electron-ion-plasma viewpoint [36].

6. Heterovalent cation impurities in molten salts

The last topic that I wish to touch upon concerns the local liquid structure around heterovalent cation impurities in molten salts. For the case of transition-metal ions this is experimentally reflected in their optical spectra through the induced splitting of the d-electron energy levels [37]. Many cases have been examined using a liquid LiCl-KCl mixture of eutectic composition as the solvent and the data have been interpreted in terms of the formation of chlorocomplexes.

For many divalent ions (Ni^{2+}, Cu^{2+}, Cr^{2+}, Fe^{2+}, Co^{2+} and Mn^{2+}) a tetrahedral-type coordination is observed, with an appreciable distortion from ideal tetrahedral symmetry for the three former ions. This observation is interesting for the following reasons:

(i) the occurrence of tetrahedral coordination for divalent 3d-transition metal impurities seems to be extremely rare in binary crystalline phases, and in particular in the crystalline alkali halides such impurities seem to enter substitutionally (coordination number 6) apart from the case of Co^{2+} which enters interstitially (coordination number 4) in both NaCl and KCl [38];

(ii) the octahedral configuration is believed [37] to be energetically favoured even in the liquid phase by "crystal field" interactions for all the above mentioned transition metal ions, with the exception of Mn^{2+} whose ground state is not split in a crystal field.

The experimental situation for trivalent transition-metal impurities in molten salts is even more complex and interesting [37]. In the case of Cr^{3+} only the octahedral-type configuration is reported, but in the cases of Ti^{3+} and V^{3+} (and also in the case of V^{2+}) the spectroscopic data show a transition occurring with increasing temperature from a preferred octahedral configuration to a situation of equilibrium between octahedral and distorted tetrahedral complexes in coexistence.

Two qualitative physical effects seem to underlie the behaviour summarized above, namely the Coulomb interactions between the heterovalent cation and its neighbours in the molten salt and the relative entropy of the two alternative configurations [39]. Simple model calculations show that, depending on its size, the heterovalent ion may gain electrostatic free energy through a tightening of its coordination shell rather than through an increase of its coordination number - indeed a tight coordination shell implies a low coordination number, on account of the size of the halogen ions which constitute this shell. These calculations predict an equilibrium coordination number of 4 in the specific case of divalent impurities with an ionic radius typical of transition-metal ions ($r_M \cong 0.6 \div 0.7$Å) dissolved in the LiCl-KCl eutectic mixture. The calculated coordination numbers for the solvent cations ($n_{Li} \cong 3.5$ and $n_K \cong 5$) are also in approximate agreement with computer simulation results [40] for this mixture. It thus seems that in the case of transition-metal ions this tendency to four-fold coordination induced by Coulomb interactions may counterbalance in many cases the "crystal field" effects favouring six-fold coordination.

An entropy difference between these two alternative coordination states, in the presence of "crystal field" effects, may also be an important factor in determining their equilibrium at high temperatures. Very rough estimates on a solid-like model of the two types of complex in the rocksalt structure suggest that the entropy of the four-fold coordination state may be appreciably larger.

7. Concluding remarks

The detailed calculations of liquid structure for molten salts that are involved in the theoretical work described in this article have been based on a charged-hard-spheres model in the mean spherical approximation. It may be useful to stress that this approach is particularly convenient in handling rather complex physical situations because of the simplicity of the basic model and of its numerical implementation. At the same time it should be stressed that this approach yields at most a semiquantitative gross picture. The usefulness of better treatments of the ionic interactions and of the statistical mechanical problem, which are practicable in at least some of the areas touched upon in this article, should not be forgotten.

Acknowledgements

This article was prepared during leave of absence from the Istituto di Fisica Teorica,

Universita di Trieste. It is pleasure to acknowledge the general support received from the Consiglio Nazionale delle Ricerche, the International Centre for Theoretical Physics, and the Scuola Internazionale Superiore di Studi Avanzati.

References

1. D.I. Page and I. Mika, J.Phys.C4, 3034 (1971).
2. F.G. Edwards, J.E. Enderby, R.A. Howe and D.I. Page, J.Phys.C8, 3483 (1975).
3. J.Y. Derrien and J.Dupuy , J. Physique 36, 191 (1975).
4. E.W. Mitchell, P.F.J. Poncet and R.J. Stewart, Phil. Mag. 34, 721 (1976).
5. L.V. Woodcock and K. Singer, Trans. Faraday Soc. 67, 12 (1971); E.M. Adams, I.R. McDonald and K. Singer, Proc. Roy. Soc. A357, 37 (1977).
6. M.C. Abramo, C. Caccamo, G. Pizzimenti, M. Parrinello and M.P. Tosi, J.Chem. Phys. 68, 2889 (1978).
7. M.J.L. Sangster and M. Dixon, Adv. Phys. 25, 247 (1976).
8. J.P. Hansen and I.R. McDonald, Phys. Rev. A11, 2111 (1975).
9. D.L. Price and J.R.D. Copley, Phys. Rev. A11, 2124 (1975); J.R.D. Copley and G. Dolling, J. Phys. C11, 1259 (1978).
10. D.G. Mead, J. Phys. C7, 445 (1974); D.G. Mead and G.R. Wilkinson, Proc. Roy. Soc. A354, 349 (1977).
11. M. Parrinello and M.P. Tosi, Riv. N. Cimento 2, no. 6 (1979); M.P. Tosi, in "Modern Trends in the Theory of Condensed Matter", p. 150 (eds. A. Pekalski and J. Przystawa; Springer Verlag, Berlin 1980).
12. F.G. Fumi and M.P. Tosi, J. Phys. Chem. Solids. 25, 31 (1964); M.P. Tosi and F.G. Fumi, J. Phys. Chem. Solids 25, 45 (1964).
13. J.P. Hansen and L. Verlet, Phys. Rev. 184, 150 (1969).
14. T.V. Ramakrishnan and M. Yussouff, Solid State Commun. 21, 389 (1977) and Phys. Rev. B19, 2775 (1979).
15. N.H. March and M.P. Tosi, Phys. Chem. Liquids 10, 185 (1980).
16. H. Reiss, S.W. Mayer and J.L. Katz, J. Chem. Phys. 35, 820 (1961).
17. N.H. March and M.P. Tosi, Phys. Chem. Liquids 11, 79 (1981).
18. F.G. Edwards, R.A. Howe, J.E. Enderby and D.I. Page, J. Phys. C11, 1053 (1978).
19. N.H. March and M.P. Tosi, Phys. Chem. Liquids 11, 89 (1981).
20. E. Mollwo, Nachr. Ges. Wiss. Göttingen 1, 203 (1935); D.M. Gruen, M. Krumpelt, and I. Johnson, in "Molten Salts: Characterization and Analysis", p. 169 (ed. G. Mamontov, Marcel Dekker, New York 1969).
21. K.S. Pitzer, J. Amer. Chem. Soc. 84, 2025 (1962).
22. N.H. Nachtrieb, Adv. Chem. Phys. 31, 465 (1975).
23. G. Senatore, M. Parrinello and M.P. Tosi, Phil. Mag. B41, 595 (1980).
24. J.F. Rounsaville and J.J. Lagowski, J. Phys. Chem. 72, 1111 (1968).
25. N.H. Nachtrieb, private communication.
26. M.A. Bredig in "Molten Salt Chemistry", p. 367 (ed. M. Blander; Interscience, New York 1964).
27. J.D. Corbett, in "Fused Salts", p. 341 (ed. B. Sündheim; McGraw-Hill, New York 1964).
28. H.R. Bronstein and M.A. Bredig, J. Amer. Chem. Soc. 82, 2077 (1958) and J. Phys. Chem. 65, 1220 (1961); H.R. Bronstein, A.S. Dworkin and M.A. Bredig, J. Chem. Phys. 37, 677 (1962).
29. I. Katz and S.A. Rice, J. Amer. Chem. Soc. 94, 4874 (1972).
30. P.J. Durham and D.A. Greenwood, Phil. Mag. 33, 427 (1976).
31. G. Senatore, P.V. Giaquinta and M.P. Tosi, Physica 112B, 360 (1982).
32. J.F. Jal, P. Chieux, J. Dupuy and J.P. Dupin, J. Physique 41, 657 (1980); P. Chieux, P. Damay, J. Dupuy and J.F. Jal, J. Phys.Chem. 84, 1211 (1980); G. Chabrier, J.F. Jal, P. Chieux and J. Dupuy , in the press.
33. J.F. Jal, Thesis at the University Claude Bernard - Lyon I (1981).
34. G. Chabrier, G. Senatore and M.P. Tosi, N. Cimento (in the press).
35. F. Hensel, Adv. Phys. 28, 555 (1979).
36. D.K. Chaturvedi, G. Senatore and M.P. Tosi, Lett. N. Cimento 30, 47 (1981); D.K. Chaturvedi, M. Rovere, G. Senatore and M.P. Tosi, Physica 111B, 11 (1981).

37. D.M. Gruen and R.L. McBeth, Pure Appl. Chem. 6, 23 (1963); D.M. Gruen in "Fused Salts", p. 301 (ed. B. Sundheim; McGraw-Hill, New York 1964).
38. S. Hirako and R. Onaka, J. Phys. Soc. Japan, 47, 1019 (1979); T. Iri, S. Hirako and K. Kambe, J. Phys. Soc. Japan, 46 106 (1979); T. Nasu, Phys. Stat. Sol. (b) 70, 97 (1975).
39. W. Andreoni, M. Rovere and M.P. Tosi, Phys. Chem. Liquids (in the press).
40. C. Caccamo and M. Dixon, J. Phys. C13, 1887 (1980).

DISCONTINUITIES OF SOME PROPERTIES OF SODIUM AMMONIA SOLUTIONS ABOVE THE NORMAL MISCIBILITY GAP.

R. Winter and U. Schindewolf, Institute of Physical Chemistry and Electrochemistry, University of Karlsruhe, Germany; A. Voronel, Department of Physics and Astronomy, University of Tel-Aviv, Israel

In our thermodynamic studies via EMF-measurements /1/ we obtained the chemical potential μ_{Na}, the partial molar enthalpy H_{Na}, entropy S_{Na}, volume V_{Na} and heat capacity $C_{p,Na}$ of the dissolved sodium in sodium-ammonia solutions in a wide concentration and temperature range. However, due to experimental reason we could not get $C_{p,Na}$ in the concentration range around the miscibility gap (T_c = -41°C; $x_c \approx$ 4 MPM). Consequently we tried to get the missing data by direct determination of the specific heat C_s of the solutions with a precision adiabatic calorimeter /2/.

The expected discontinuities of C_s were observed leaving by temperature increase the normal miscibility gap. But in addition further jumps of the C_s/T-curves were found at $T > T_c$ and $x > x_c$ which we interpreted as another region of instability. To confirm this careful measurements of the EMF /3/ and of the electrical resistivity ϱ of the solutions were carried out in the interesting T- and x-range. The electrochemical cell Ag / Na // $Na^+ .... e^-$ / Ag with the Na^+-conducting Na-ß-alumina (symbolised by //) responds reversibly to dissolved sodium ($Na^+ + e^-$); thus the EMF and its temperature dependence give the chemical potential μ_{Na} and the partial molar entropy S_{Na} of the dissolved sodium relative to those of solid sodium.

The obtained data (Fig. 1 and 2 as examples) also reveal sudden charac-

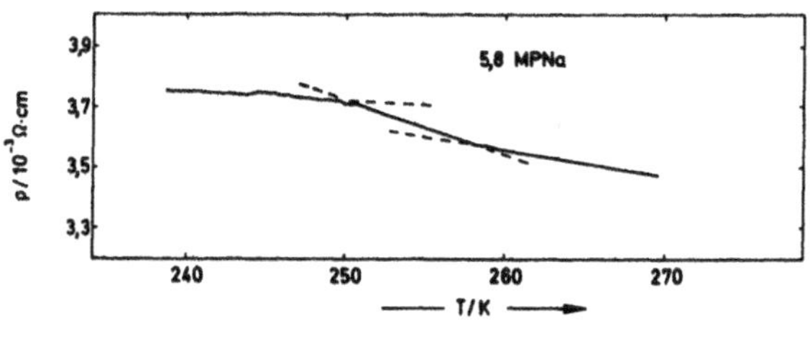

Fig. 1. Continous registration of the temperature dependence of the specific resistivity of a 5.8 MPM Na-NH_3 solution.

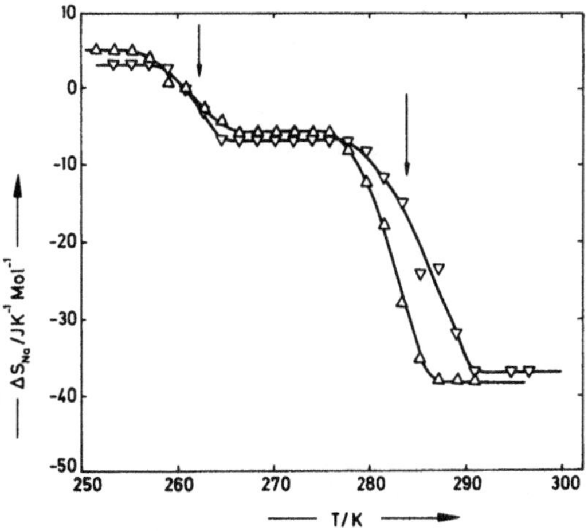

Fig. 2. Temperatur dependence of the partial molar entropy of Na in a 8.5 MPM Na-NH_3 solution (relative to that of solid Na).

teristic changes of the S_{Na}/T and ϱ/T-curves within the same x- and T-range as found by measurements of the heat capacity. In Fig. 3 the T-x-data are combined at which discontinuities of C_s, S_{Na} or ϱ were observed. It is obvious that all three measurements lead to the same results which identify another region of instability above the normal miscibility gap.

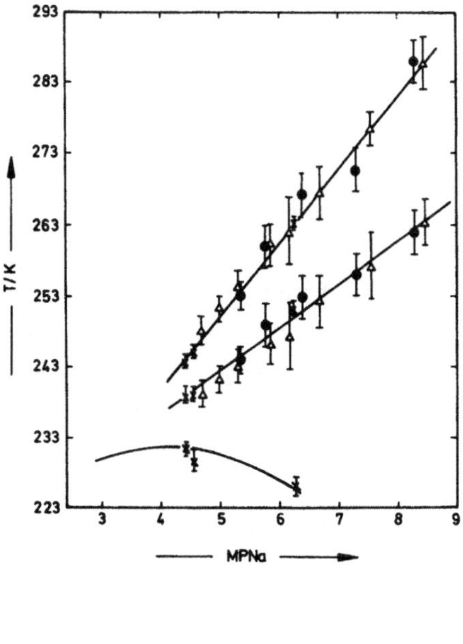

Fig. 3. Phase diagram of Na-NH_3 solutions. The experimental points indicate sudden changes of the heat capacity (x), of the partial molar entropy of Na (Δ) and of the electric conductivity (●). On the lower left hand the normal miscibility gap is indicated.

In the above experiments the formation of a solid compound was observed on Pt-electrodes in the solutions of the T- and x-range of the instability. Outside of it or on Au, Ag, W, Mo and other metals no solid precipitate is formed. The solid has the composition $Na(NH_3)_3$, in else it is an amoniate. It has a simple cubic structure with a lattice constant of 5.1 Å (Fig. 4). Its conductivity behaviour is that of a semiconductor. The compound is diamagnetic (probably antiferromagnetic) with a slight paramagnetic contribution. With increasing temperature the ESR-spectrum changes at -80°C reversibly from a wide (half width 3 to 6 G) to a narrow line (half width .7G at 9.4 GHz), revealing a magnetic transition. The compound decomposes above 0°C; below the decomposition temperature ammonia can be pumped off reversibly, leaving metallic sodium in a spongy form.

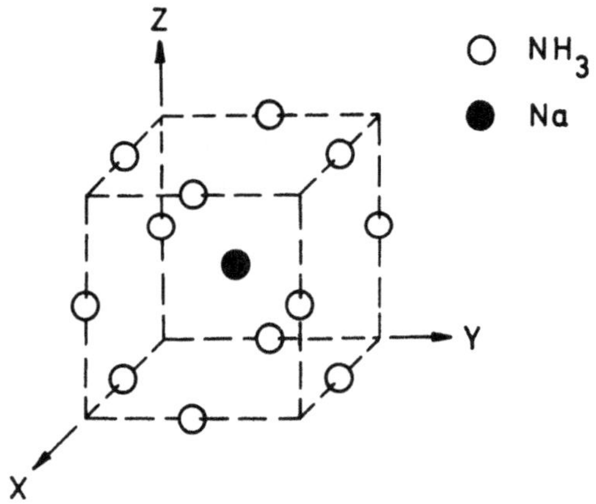

Fig. 4. Proposed crystal structure of $Na(NH_3)_3$; lattice constant 5.1 Å.

Finally in the same T-x-range of the phase instability P. Arendt, Berlin /4/, observed sudden changes of the electrical conductivity of sodium-ammonia solutions in contact with air. Dmitrenko, Charkow, and we could confirm this strange behaviour. The current-voltage relation changes with time indicating increasing resistance of the solution due to partial decomposition (Fig. 5). Then after a few minutes a drastic change of the resistivity occurs: very high currents are observed with a very small voltage drop, indicating an increase in conductivity by a factor of up to ten thousand.

Presently we refrain from giving any interpretation of the observed effects until more experimental data are available. But it seems to us that the three effects are somehow related, because all of them are observed only in the same T-x-range. And we feel that the observed instability (Fig. 3) is the first experimental proof of the general theory /5,6/ predicting two critical points in ionised or metallic systems.

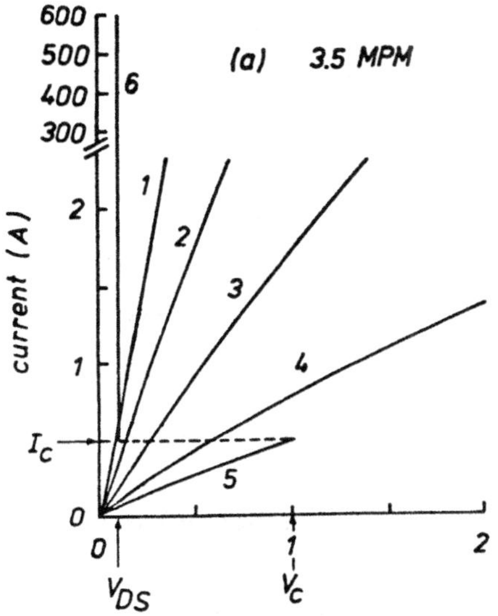

Fig. 5. Current-voltage relation of a 3.5 MPM Na-NH_3 solution at the normal boiling point in contact with air. Curves 1 to 6 were obtained in succeeding runs within a few minutes /4/.

We are grateful to Profs. H. Bärnighausen (Karlsruhe), F. Baumann (Karlsruhe), J. Dye (East Lansing) and W. Westerhaus (Stuttgart) for cooperation and help in obtaining the x-ray diffraction pattern, the data on the electrical conductivity and on the magnetic properties of the described solid compound. We acknowledge the financial support of the Deutsche Forschungsgemeinschaft and the gift of Na-ß-alumina by BBC (Heidelberg).

1) U.Schindewolf and M.Werner,J.Phys.Chem. 84,1123 (1980).

2) V.Steinberg, A.Voronel, D.Linsky and U.Schindewolf, Phys.Rev.Let. 45, 1338 (1980).

3) R.Winter, U.Schindewolf and A.Voronel, J.Phys.F; Metal Phys.11, 281 (1981).

4) P.Arendt, private communication.

5) W.Ebeling and R.Sändig, Ann.Phys. 28, 289 (1973).

6) M.Gitterman and V.Steinberg, Phys.Rev.A, 20, 1236 (1979).

THE SURFACE PROPERTIES OF MOLTEN SALTS

R. Evans

H.H. Wills Physics Laboratory,
University of Bristol,
Bristol, BS8 1TL,
U.K.

ABSTRACT

We review recent theoretical work on the structure and surface tension of the liquid-vapour interface of molten salts.

During the last few years there has been rapid progress in the theory of the surface tension and equilibrium structure of the liquid-vapour interface of pure, simple (argon-like) fluids [1,2]. Tractable statistical theories, which allow one to calculate the density profiles, the degree and sign of the relative adsorption, and the surface tension from a knowledge of the interatomic potentials have also been developed for binary mixtures of simple fluids [3,4]. By contrast the surfaces of molten salts have received considerably less attention. This is somewhat surprising when it is realized that the liquid-vapour interface of such liquids can be viewed as one of the simplest electrochemical systems. At low temperatures the density profiles of the anions and the cations will vary extremely rapidly at the interface so that the system constitutes a highly inhomogeneous charged fluid.

The density functional approach [1,2] which has proved successful for describing the interfaces of simple fluids, where the interatomic potential is short-ranged, can be extended [5] to systems where the interparticle potentials are Coulombic for large separations. Approximation schemes [5] have been developed for calculating the average one-particle density profiles and the interfacial thermodynamic properties of ionic liquids; such schemes represent systematic improvements on the Gouy-Chapman or Debye-Hückel theories commonly used in electrolyte theory. In this lecture we concentrate on the application of the theory to the liquid-vapour interface of models of molten alkali-halides.

Results of calculations [6], based on a square-gradient approximation to the Helmholtz free-energy functional, of the density profile and surface tension of the restricted primitive model of a molten salt are presented. In this model the anion and cation have equal sizes and there is no electrical double layer at the liquid-vapour interface. For low temperatures, the density profile has a different shape from that calculated for Lennard-Jones fluids. The results for the interfacial width are in reasonable agreement with the results of a computer simulation [7] of a model of molten KCl; the interfacial width is approximately one hard-sphere diameter near the melting temperature. Our calculated surface tensions are in qualitative agreement with the results of the simulation and with experimental data for those alkali halides where the anion and cation are of similar size. Our theory provides an explanation of why the surface tensions of molten alkali halides are relatively small - given the large binding energies and melting temperatures of these materials.

When the anion and cation differ in size an electrical double layer develops at the liquid-vapour interface. The magnitude and sign of the change in electostatic potential across the interface and its effect on the surface tension depends upon the particular alkali halide under consideration [8]. In general the larger species prefers to sit in the vapour side of the interface [8].

We will comment briefly on the form of pairwise correlations at the liquid-vapour interface of a salt and on the form of the charge-charge correlation function for a molten salt near a wall.

Finally we mention recent theoretical work [9] on the surface properties of liquid alkali metals. These can also be considered as inhomogeneous ionic liquids since the ions and the conductions form a weakly-coupled electron-ion plasma.

REFERENCES

1. R. Evans, 1979, Adv. in Phys., 28, 143.
2. F.F. Abraham, 1979, Phys. Rep., 53, 93.
3. M.M. Telo da Gama and R. Evans, 1980, Molec. Phys., 41, 1091.
4. M.M. Telo da Gama and R. Evans, 1982, Faraday Symp. No. 16, to be published in Discuss. Faraday Soc.
5. R. Evans and T.J. Sluckin, 1980, Molec. Phys., 40, 413.
6. M.M. Telo da Gama, R. Evans and T.J. Sluckin, 1980, Molec. Phys., 41, 1355.
7. D.M. Heyes and J.H.R. Clarke, 1979, J. Chem. Soc. Faraday II, 75, 1240.
8. T.J. Sluckin, 1981, J. Chem. Soc. Faraday II, 77, 1029.
9. R. Evans and M. Hasegawa, 1981, J. Phys. C: Sol. St. Phys., 14, 5225.

CHARGE ORDERING IN MOLTEN SALTS AND IN SOME LIQUID SEMICONDUCTING ALLOYS.

A.P. Copestake and R. Evans

H.H. Wills Physics Laboratory,
University of Bristol,
Bristol BS8 1TL, U.K.

ABSTRACT

We summarize the main results of our study of 'charge-ordering' in charged-hard sphere models of molten salts and liquid semiconductors. We demonstrate that provided the attraction between unlike species is strong, very short-ranged screened interionic potentials can produce partial structure factors and radial distribution functions which are very close to those found in molten alkali halides. It is the behaviour of the charge-charge or, equivalently, the concentration-concentration structure factor $S_{cc}(k)$ at small k which provides the most direct information about the strength and range of the interionic potentials. We discuss, briefly, a calculation of $S_{cc}(k)$ in liquid Li_4Pb.

It is well-known that in molten alkali-halides the partial radial distribution functions $g_{ij}(r)$ exhibit 'charge-ordering'; $g_{+-}(r)$ oscillates out of phase with $g_{++}(r)$ and $g_{--}(r)$ to give pronounced oscillations in the charge-charge radial distribution function $g_{QQ}(r) = \frac{1}{2}(g_{++}(r) + g_{--}(r) - 2g_{+-}(r))$. In reciprocal space this 'charge-ordering' gives rise to partial structure factors $S_{++}(k)$ and $S_{--}(k)$ whose main peaks lie at almost the same value of k as the position of the first minimum of $S_{+-}(k)$ [1,2]. Since statistical theories and computer simulations [2] have shown that models of molten salts in which charged hard-spheres interact via Coulomb potentials give a rather good account of the measured structure factors and of the oscillatory behaviour of $g_{QQ}(r)$ of the alkali halides, it is often argued that the 'charge-ordering' found in real salts is a signature of Coulombic inter-ionic potentials.Consequently if diffraction experiments on a binary liquid indicate some 'charge-ordering' this might be taken as evidence for some kind of ionic bonding in the particular liquid under consideration [3,4]. Attempting to infer the nature of the interatomic potentials in a binary liquid from structural data alone is, of course, a difficult and, perhaps, unrewarding task so before embarking on sophisticated model building and analysis of diffraction data for specific liquids it is natural to ask to what extent do the $S_{ij}(k)$ and the $g_{ij}(r)$ yield detailed information about the interatomic potentials? Could one decide unambiguously that a particular binary liquid was ionic, say, if the only experimental data one had was structure factor data? Can very different forms of interatomic potentials lead to the same or very similar $S_{ij}(k)$ and $g_{ij}(r)$?

These questions have particular relevance for liquid semiconducting alloys [5] and for those metallic alloys where there is pronounced chemical short range order [6]. Because of the complexity of the electronic structure of these alloys even the qualitative form of the bonding is often uncertain and structural data could play a useful role in elucidating characteristic features [5,3]. In the present study we focus attention on those liquid semiconducting alloys where there is reason to believe that pronounced charge transfer occurs between the species leading to what can be described loosely as partial ionic bonding. The Au-alkali metal alloys fall into this class. While there is much evidence [7] to suggest that equimolar AuCs and AuRb are fully ionized, i.e. they consist of Au^- and Cs^+ or Rb^+ ions, so that the interatomic potentials and, therefore, the structure factors of these alloys should be similar to those of the molten alkali halides [8,3], the degree of charge transfer and the precise form of the interatomic potentials in the other Au-alkali alloys is not fully understood [7,9]. Liquid alloys of Li with the polyvalent metals Tℓ, Pb, Sn, Sb and Bi also appear to exhibit pronounced charge transfer. Evidence for this comes from measurements of electrical transport properties [10,11], spin-flip cross-sections [10], the Knight shift [12] and magnetic susceptibility [13]. We might suppose that in these Li alloys and in Au(K, Na or Li) the interatomic potentials, at the appropriate stoichiometric composition, could be modelled in a zeroth approximation by hard-spheres with screened Coulomb tails. Unlike species would interact via an attractive screened potential and like species via repulsive screened potentials outside the hard-cores. The strengths of these potentials would depend on the extent of the charge transfer and the ranges on the conduction electron density; most of these alloys have nearly-metallic conductivities and some are (poor) metals. (A model of this type was employed by Schirmacher [14] in calculations of the thermodynamic properties of Li_4Pb.) One might expect the partial structure factors and radial distribution functions of such models to differ from those appropriate to molten salts, i.e. from those obtained from hard-spheres with Coulomb tails. In particular one might guess that the 'charge-ordering' should be less-pronounced when the potentials are screened.

In an attempt to understand the structure of this type of liquid semiconducting alloy and, more specifically, to investigate 'charge' fluctuations, we have calculated [15] $g_{ij}(r)$ and $S_{ij}(k)$ for two different restricted primitive, charged hard-sphere models. In the first the pairwise potentials are taken to be

Coulombic outside the core:

$$\phi_{ij}^{C}(r) = \begin{cases} \infty & r < d \\ e_i e_j / \varepsilon r & r > d \end{cases} \quad (1)$$

where d is the hard-sphere diameter (assumed the same for each species), ε is the dielectric constant of the medium in which the ions are immersed and e_i is the charge on species i. We have set $|e_i| = |e_j| = e$ and imposed overall charge neutrality. The state we have considered has a reduced number density $\rho^* = Nd^3/V = 0.669$, where N is the total number of ions and V is the total volume, and a reduced inverse temperature $\beta_C^* = e^2/\varepsilon d k_B T = 35.476$. If we set $\varepsilon = 1$, e = electron charge and d = 3Å, say, the model would be appropriate for a molten alkali -halide at about 1500K. The second model takes screened Coulomb (Yukawa) potentials outside the core:

$$\phi_{ij}^{Y}(r) = \begin{cases} \infty & r < d \\ \dfrac{e_i e_j}{\varepsilon r} \exp\left[-z(r/d - 1)\right] & r > d \end{cases} \quad (2)$$

where z is a dimensionless inverse screening length. Following an earlier investigation of this model by Larsen and Rogde [16], calculations were performed for the state with the same density as above but with $\beta_Y^* = 5.3197$ and $z = 4$. With this choice of parameters the potentials defined by (1) and (2) differ enormously for $r > d$. At $r = d$ the reduced Yukawa potentials $\phi_{ij}^{Y}/K_B T$ are a factor of $\beta_C^*/\beta_Y^* = 6.67$ smaller in magnitude than the corresponding reduced Coulomb potentials $\phi_{ij}^{C}/K_B T$ and they decay almost to zero for $r \sim 2d$. In spite of the very large differences between the potentials our calculations [15], which are based on the mean-spherical-approximation (MSA) and the hyper-netted-chain (HNC) approximation, yield surprisingly similar results for $g_{ij}(r)$ and $S_{ij}(k)$. The radial distribution functions for the Yukawa model exhibit 'charge-ordering'; $g_{++}(r) \equiv g_{--}(r)$ oscillates out of phase with $g_{+-}(r)$ and the first peak in $S_{++}(k) \equiv S_{--}(k)$ lines up with the first minimum in $S_{+-}(k)$. 'Charge-ordering' is not a unique feature of Coulombic interactions but seems to occur whenever the potential between unlike species is strongly attractive. There are no features in the $g_{ij}(r)$ which one could identify as being especially characteristic of Coulombic potentials or which provide a useful signature of Coulombic potentials - a point made first by Larsen and Rogde [16]. Debye-Hückel types of argument provide a qualitative explanation of the similarity between the results for $g_{ij}(r)$ in the two models [16,15].

Our analysis of the partial structure factors suggests that results in k-space shed more light on the form of the interionic potential - especially when these are expressed in terms of number, N, and charge, Q, variables or, equivalently, number and concentration, C, variables. For each model the number-number structure factor $S_{NN}(k) = S_{++}(k) + S_{+-}(k)$ is dominated by the hard-sphere contribution so it is evident that this function will be almost the same for both cases. Indeed in the MSA $S_{NN}(k)$ is identical to the Percus-Yevick hard-sphere structure factor for both models. The number-charge structure factor $S_{NQ}(k) = S_{++}(k) - S_{--}(k)$ is identically zero for the present models. It is only the charge-charge structure factor $S_{QQ}(k) = S_{++}(k) - S_{+-}(k)$ which usefully characterizes the interionic potentials. The Coulomb potential yields a higher first maximum in $S_{QQ}(k)$ than the Yukawa potential and this reflects somewhat stronger 'charge-ordering' in $g_{QQ}(r)$. The positions of the first maxima are, however, almost identical and for larger k the Yukawa and Coulomb results are virtually indistinguishable. It is only for k in the range $0 \leqslant k \lesssim 1\text{Å}^{-1}$, that significant differences are found. For the Coulombic model it is well-known that

$$S_{QQ}(k) = k^2/k_D^2 + O(k^4) \tag{3}$$

where k_D, the inverse Debye screening length, is given by $k_D^2 = 4\pi e^2 N/V\varepsilon k_B T$. This result is only valid for interionic potentials which vary asymptotically as 1/r and it is a statement of the Stillinger-Lovett [17] moment conditions. For the Yukawa model, on the other hand, $S_{QQ}(0) \neq 0$ since there is no requirement that long-wavelength fluctuations in the total 'charge' density must vanish when the interactions are screened from the outset. Our calculations indicate that $S_{QQ}(0)$ is small for the Yukawa model so that this binary liquid is also highly ordered; long-wavelength 'charge', or more accurately, concentration fluctuations are surpressed. If we expand $S_{QQ}(k)$ for small k:

$$S_{QQ}(k) = S_{QQ}(0) + ak^2 + O(k^4) \tag{4}$$

we find [15] that the coefficients $S_{QQ}(0)$ and a depend rather sensitively on the range, z^{-1}, and the strength, β_Y^*, of the Yukawa potential. $S_{QQ}(0)$ decreases with increasing z^{-1} and increasing β_Y^*; increasing the range and strength of the ordering potential surpresses further the long-wavelength concentration fluctuations. The height of the first maximum in $S_{QQ}(k)$ also depends on the range and strength of the ordering potential $\phi_{++}(r) - \phi_{+-}(r)$ but in a more complex fashion.

We conclude from these results, and from results of calculations for other states, that provided the Yukawa potentials are sufficiently strong, short-ranged interactions can produce <u>short-range order</u> which is very similar to that found in molten salts. It is only at small k where the different behaviour of the interionic potentials will manifest itself in a direct way. This suggests that low angle diffraction experiments, which can in favourable circumstances yield $S_{QQ}(k)$, or more accurately, $S_{cc}(k)$, might provide useful information concerning the range and strength of the interionic potentials.

As an application of the screened Coulomb model we have performed calculations of the partial structure factors and radial distribution functions for the alloy Li_4Pb. (This work, which was performed in collaboration with H. Ruppersberg and W. Schirmacher, will be reported in detail elsewhere). Li_4Pb is of particular interest since there is neutron diffraction data for $S_{cc}(k)$ in this system [18,19]. We have used model (2) for the interatomic potentials with $\varepsilon = 1$ and the hard-sphere diameter $d = 2.65$Å, corresponding to the effective diameter of both Li and Pb in the alloy. The inverse screening length $\kappa = z/d$ was set equal to 1.1Å^{-1} and the charges on the Li and Pb were varied - but under the electroneutrality restriction that $\sum_{i=1,2} \rho_i e_i = 0$, where ρ_i is the number density of species i. We find that for small charge transfers, typically +0.12e or +0.15e on a Li (-0.48e or -0.6e on a Pb) the results of our MSA and HNC calculations are in good qualitative agreement with the experimental results for $S_{cc}(k)$ and the Fourier transform of this quantity, $g_{cc}(r)$. It appears that our simple model gives an adequate description of the near salt-like 'charge ordering' which occurs in this alloy [18,19]. Further work is required in order to determine its usefulness for quantitative purposes and its applicability as a reference system for describing the structure of other liquid semiconducting alloys. In this context we mention that the HNC can be applied readily to 'soft' core potentials and to binary liquids where the species have different diameters.

REFERENCES

1. J.E. Enderby and G.W. Neilson, 1980, Adv. Phys., 29, 323.
2. M. Parrinello and M.P. Tosi, 1979, Riv. Nuovo Cim., 2, Nr.6.pg. 1.
3. R. Evans and M.M. Telo da Gama, 1980, Phil. Mag., 41, 351.
4. A.P. Copestake, R. Evans and M.M. Telo da Gama, 1980, J. de Physique, 41, Coll. 8, 321.
5. J.E. Enderby, 1978, in 'The Metal Non-Metal Transition in Disordered Systems', eds. L.R. Friedman and D.P. Tunstall (University of Edinburgh Press), pg. 425.
6. P. Chieux and H. Ruppersberg, 1980, J. de Physique, 41, Coll. 8, 145.
7. F. Hensel, 1979, Adv. Phys., 28, 555.
8. W. Martin, W. Freyland, P. Lamparter and S. Steeb, 1980, Phys. Chem. Liquids, 10, 61.
9. Theories for charge-transfer and the electronic structure of Au-alkali alloys have been developed - see C. Holzhey, F. Brouers and J.R. Franz, 1981, J. Phys. F, 11, 1047 and references therein.
10. V.T. Nguyen and J.E. Enderby, 1977, Phil. Mag., 35, 1013.
11. C. van der Marel, A.B. van Oosten, W. Geertsma and W. van der Lugt, J. Phys. F (to appear).
12. C. van der Marel, W. Geertsma and W. van der Lugt, 1980, J. Phys. F., 10, 2305.

13. K. Hackstein, S. Sotier and E. Lüscher, 1980, J. de Physique, 41, Coll.8, 49.
14. W. Schirmacher, 1980, J. Non-Cryst. Solids 35 and 36, 1301.
15. A.P. Copestake and R. Evans, 1982, J. Phys. C., (to appear).
16. B. Larsen and S.A. Rogde, 1980, J. Chem. Phys., 72, 2578.
17. F.H. Stillinger and R. Lovett, 1968, J. Chem. Phys., 48, 3858. Ibid. 49, 1991.
18. H. Ruppersberg and H. Egger, 1975, J. Chem. Phys., 63, 4095.
19. H. Ruppersberg and H. Reiter, 1982, J. Phys. C. (to appear).

HYPERSONIC RELAXATION IN A GLASS FORMING KNO_3-$Ca(NO_3)_2$ MIXTURE

L.M. Torell
Department of Physics, Chalmers University of Technology,
S-412 96 Gothenburg, Sweden

ABSTRACT

Brillouin scattering (GHz) experiments were performed on a viscous 60 % KNO_3 - 40 % $Ca((NO_3)_2$ mixture (mol. %) in a temperature range of 90-380°C. Sound velocities for both the transverse and longitudinal waves were measured as well as the absorption. A single relaxation time behaviour was revealed for temperatures > 200°C. The temperature dependence of the relaxation time was determined in a wide range, ~300°, above the glass transition.

I. INTRODUCTION

In this laboratory a transverse mode due to oscillatory shear waves has recently been observed for the first time in a Brillouin scattering spectrum of a molten salt[1]. This became possible through using a nitrate mixture of K-Ca which is capable of extensive supercooling to a high viscosity liquid and to a glass. The K-Ca nitrate system has widely been studied by a variety of techniques over the composition range 25-50 mol % $Ca(NO_3)_2$ where it is glassforming, among these are the viscosity[2-4] and ultrasonic[5] studies of special interest for the present work. Measurements of the transverse[3,4] and longitudinal[5] sound velocity have been reported in the kHz and MHz frequency range, where a structural relaxation process was revealed with a broad distribution of relaxation times at low temperatures. For higher temperatures the distribution sharpens and a single relaxation time behaviour was suggested at high enough temperatures. Similar behaviour has also been reported for the molten oxides[6-8] and $ZnCl_2$[9]. However, relaxation effects in the high temperature region could not be observed by ultrasonic technique since a shorter time process was needed. We report here a high frequency (GHz) Brillouin scattering study of structural and shear relaxation of a 60 % KNO_3-40 % $Ca(NO_3)_2$ mixture over a wide temperature range above the glass transition.

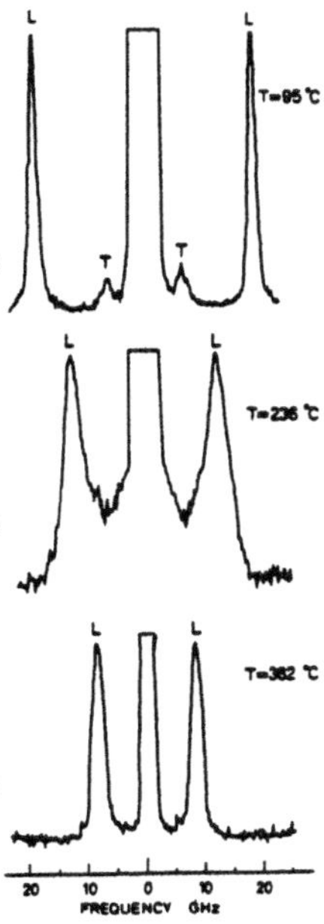

Fig. 1: Brillouin spectra at 95°, 236° and 382°C

The Brillouin scattering spectra consist of the Rayleigh line (R) centered at the frequency of the light source and a frequency shifted doublet. For amorphous solids there are two of these doublets because both shear and compressional restoring forces are present, which are responsible for transverse (T) and longitudinal (L) phonons, see Fig. 1. The transverse peaks are usually weak in intensity, about 10 to 100 times smaller than the longitudinal peaks, which in turn are orders-of-magnitudes weaker than the central line, whose intensity is governed by static inhomogeneities in the medium. The shifts ν_L and ν_T are related to the longitudinal and transverse sound velocities in the material, v_L and v_T, by

$$\nu_L = (2\, n v_L/\lambda_o)\, \sin\theta/2 \qquad (1)$$

$$\nu_T = (2\, n v_T/\lambda_o)\, \sin\theta/2 \qquad (2)$$

where θ is the scattering angle, λ_o the wavelength of the incident light, and n is the refractive index of the medium.

The melt is characterized by the ability to be compressed and sheared as longitudinal and transverse sound waves move through the fluid. At sufficiently low frequencies the ionic configurations of the liquid structure can change during the passage of the propagating pressure wave and the liquid flows and obeys Newton's viscosity law. At higher frequencies, when the time necessary to reach the new equilibrium state is comparable to the period of the perturbation, energy will be lost in the medium and relaxation phenomena will occur. This will lead to longitudinal velocity dispersion and absorption, which for a single relaxation time behaviour are expressed by[10]

$$\frac{v_L^2 - v_o^2}{v_\infty^2 - v_o^2} = \frac{\omega^2 \tau_L^2}{1 + \omega^2 \tau_L^2} \tag{3}$$

$$\alpha = \frac{v_\infty^2 - v_o^2}{2v_L^3} \frac{\omega^2 \tau_L}{1 + \omega^2 \tau_L^2} \tag{4}$$

where v_o and v_∞ are low and high frequency limits of the velocity, respectively, and α is the absorption coefficient related to the broadening of the Brillouin line Γ_β according to

$$\Gamma_\beta = \alpha v_L / 2\pi . \tag{5}$$

Combining Eqs. (3) and (4) we get an expression for the longitudinal relaxation time τ_L

$$\tau_L = \frac{1}{2\alpha} \frac{v_L^2 - v_o^2}{v_L^3} \tag{6}$$

which makes it possible to calculate τ_L from two independent measurements - the absorption and the velocity dispersion.

At high frequencies the liquid structure changes too slowly to follow the perturbation and therefore exhibits the elastic properties of a glassy solid, characterized by its limiting high frequency shear velocity $v_{T,\infty}$ and modulus G_∞ where

$$G_\infty = \rho v_{T,\infty}^2 . \tag{7}$$

G_∞ is related to the average shear relaxation time by

$$\langle \tau_s \rangle = \eta_s / G_\infty \tag{8}$$

where η_s is the shear viscosity.

II. EXPERIMENTAL METHOD

The experimental equipment is shown in Fig. 2. The light source is a single mode Ar^+ laser operating at 488.0 nm. Two scattering angles were used, 90° and 140°, to study phonons of different frequencies. The light was focussed into the sample cell situated inside a thermostat. The scattered light was analyzed by a triple-passed piezo-electrically scanned Fabry-Perot interferometer (Burleigh Instruments Inc. DAS 10) with an advanced electronic feedback system which adjusted for any non-linearities of the piezo elements during a scan. The Fabry-Perot interferometer was adjusted for a number of spectral free ranges, each chosen to get the Brillouin peaks as well resolved from the Rayleigh line as possible at that particular temperature. Overlapping orders of the longitudinal modes were used when the transverse peaks were studied. The detector system includes a cooled ITT-FW 130 photomultiplier, photon counting equipment and a multichannel analyzer. For each temperature the spectrum is the result of about one thousand sweeps. The thermostat and the preparation of the sample have been described elsewhere[11,1].

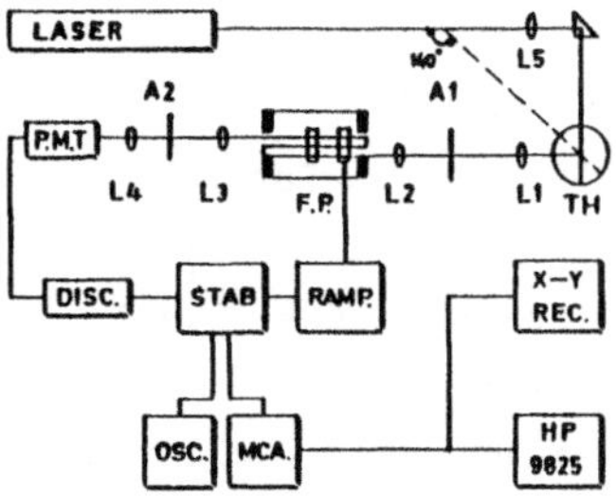

Fig. 2: The experimental equipment

III. RESULTS AND DISCUSSION

Brillouin spectra were recorded from the glass transition, 60^oC, to 382^oC. Spectra obtained at different temperatures are shown in Fig. 1, where the decreasing frequency shift with increasing temperature is for the longitudinal mode clearly seen, as well as the increasing and later decreasing line width. Also, for low temperatures where the shear viscosity is large, the weak transverse mode due to propagating shear waves was revealed.

Values of the acoustic velocities were determined from the observed frequency shifts, $\Delta\nu$, using Eqs. (1) and (2). Refractive indices used in the calculations were taken from Ref. 12. The derived data for the longitudinal velocity are plotted versus temperature for two scattering angles, Fig. 3, along with earlier ultrasonic results[5]. The complete dispersion curve was revealed and present hypersonic data are in excellent agreement with reported ultrasonic results at high and low temperatures. However, it can immediately be seen that the reported temperature dependence of $v_{L,\infty}$ is too large.

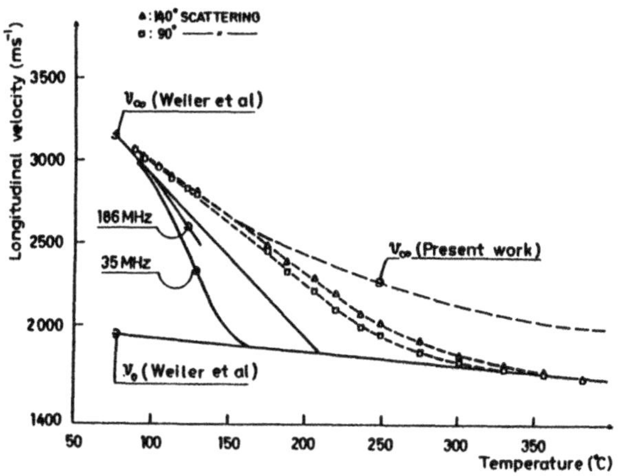

Fig. 3: Velocity versus temperature. Solid lines are from ultrasonic data. Dashed lines represent present hypersonic results.

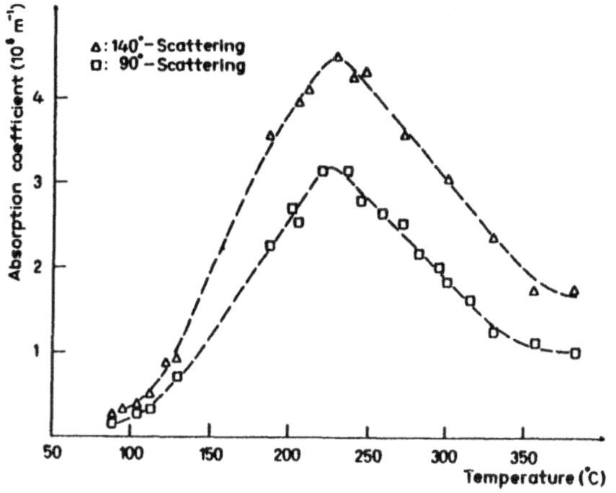

Fig. 4: Absorption coefficient versus temperature for hypersonic frequencies.

In Fig. 4 the absorption coefficient, calculated from the observed line width and Eq. (5), is plotted versus temperature. Since the absorption maximum, α_{max}, is observed, a value for the longitudinal relaxation time, τ, can directly be obtained from the peak absorbtion at $\omega\tau = 1$. For 90°-scattering the frequency of the hypersonic wave close to α_{max} is 8.91 GHz. Thus, at 220^oC, $\tau = 1.79 \times 10^{-11}$ s. For 140°-scattering the corresponding values are 11.96 GHz, 230^oC and $\tau = 1.33 \times 10^{-11}$ s. From ultrasonic values for v_o[5] and by use of Eq. (3) $v_{L,\infty}$ can now be determined for 220 and 230^oC, which combined with the present dispersionfree data at 88^oC give a mean value of the temperature coefficient, dv_∞/dt, of $-5.32\ m\ s^{-1}\ {}^oC^{-1}$. This is somewhat smaller than the directly measured values, however, a linear temperature dependence is not the appropriate form for $v_{L,\infty}$. Rather as for shear waves, it is the inverse of the limiting high frequency modulus, $1/M_\infty$, which should vary linearly with temperature[3]. The high frequency modulus is related to $v_{L,\infty}$ by

$$M_\infty = \rho v_{L,\infty}^2 \qquad (9)$$

which using a combination of the calculated value of $v_{L,\infty}$ at 220° and the measured value at 88° gives

$$1/M_\infty = (2.41 \text{ x } 2.860 \text{ x } 10^{-2}T) \text{ x } 10^{-11} \text{ N}^{-1} \text{ m}^2, \text{ (T in } ^\circ\text{C)} \qquad (10)$$

Density data were taken from Ref. 13. The corresponding curvelinear temperature dependence for $v_{L,\infty}$ is indicated by the dashed line in Fig. 3. Extrapolations of Eq. (9) now provide $v_{L,\infty}$ over the entire temperature range.

The longitudinal relaxation times at different temperatures in the interval 200° to 330°C have been calculated by using Eq. (6) and present experimental values for the absorption and velocity. The logarithmic plot of τ_L vs. the reciprocal absolute temperature is seen in Fig. 5. The best fit was obtained by choosing

$$\tau_L = 1.72 \text{x } 10^{-15} \exp(37820/RT), \quad 200 < T < 330^\circ C \qquad (11)$$

where R is the universal gas constant (J mole^{-1} K^{-1}). The standard deviations in τ_L are less than 14 %. The Arrhenius equation for the relaxation time is in accordance with the model proposed by Weiler et. al.[5]. The latter authours suggest that the behaviour of τ_L is a direct consequence of the temperature dependence of the activation energy E. At high temperatures T > 200°C, the activation energy approaches a constant value and a single relaxation is expected. The present analyses gives a constant value of the activation energy of 37.8 kJ/mole or 9.04 kcal/mole in good agreement with viscosity[2] and conductivity[14] data and any deviation from Arrhenius behaviour is beyond the accuracy of the measurements.

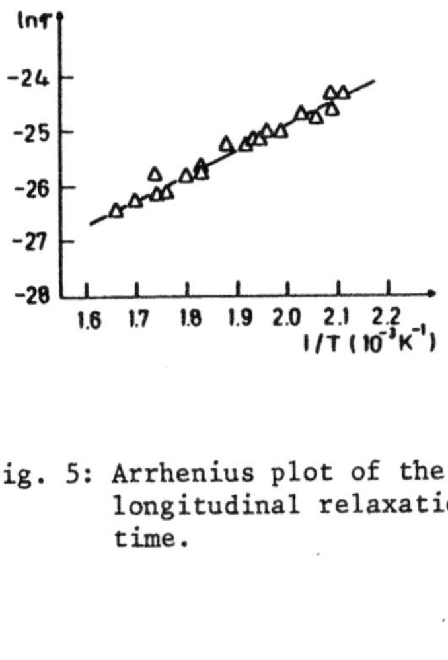

Fig. 5: Arrhenius plot of the longitudinal relaxation time.

The shear relaxation time was calculated from the transverse velocity, measured in a non-dispersion region from the glass transition to 90°C, where it may be considered as identical to $v_{T,\infty}$. The corresponding high frequency modulus G_∞ was calculated from Eq. (7) which gives

$$1/G_\infty = -3.290 \text{ x } 10^{-10} + 1.608 \text{x } 10^{-12} \text{ T (K) N}^{-1} \text{ m}^2 \qquad (12)$$

as the best fit with a standard deviation of less than 1.5 %. Extrapolations of G_∞ now provides the shear relaxation time over a wide temperature range by using Eqs. (12) and (8). In Fig. 6 present results for both the structural and the shear relaxation times are shown. For comparison data reprted for the average conductivity relaxation time[15] and a value for the NO_3^- reorientation time[16,17] have been included. Also shown are reported IR results for the inverse of the transverse vibrational mode for the quasilattice of the glass[18,19]. Fig. 6 demonstrates a natural extrapolation to 1/T = 0

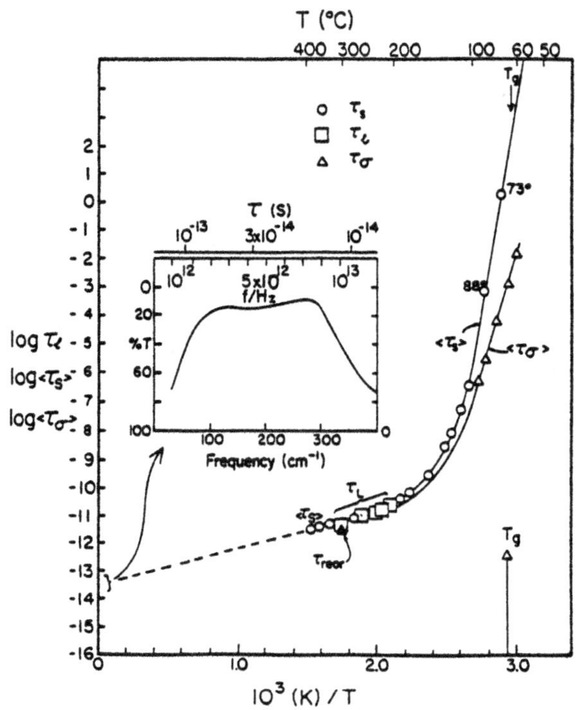

Fig. 6: Temperature dependence of the relaxation time. Present results for the shear (τ_s) and longitudinal (τ_L) relaxation times are extrapolated (dashed line) to the inverse vibrational frequency of the glass (inset). τ_{reor} is the NO_3^- reorientation time and τ_σ the conductivity relaxation time.

which carries τ directly into the region of the inverse quasi-lattice frequencies and thus determines the lower limit on τ. However, at the low temperature end τ is unbounded and the remaining question is how the **very** long τ is approached near the glass transition temperature.

REFERENCES

1. L.M. Torell, J. Chem. Phys. 76, 3467 (1982)
2. E. Rhodes, W.E. Smith, and A.R. Ubbelohde, Trans. Far. Soc. 63, 1943 (1967)
3. G.M. Glover and A.J. Matheson, Trans. Far. Soc. 67, 1960 (1971)
4. H. Tweer, N. Laberge, and P.B. Macedo, J. Am. Ceram. Soc. 54, 121 (1971)
5. R.A. Weiler, R. Bose, and P.B. Macedo, J. Chem. Phys. 53, 1258 (1970)
6. P.B. Macedo, J.H. Simmons and W. Haller, Phys. Chem. Glasses 9, 156 (1968)
7. J. Tanke, T.A. Litovitz and P.B. Macedo, J. Am. Ceram. Soc. 51, 158 (1968)
8. P.B. Macedo, J.H. Simmons and W. Haller, Phys. Chem. Glasses 9, 156 (1968)
9. G. Gruber and T.A. Litovitz, J. Chem. Phys. 40, 13 (1964)
10. T.A. Litovitz and C.M. Davis, "Physical Acoustics", Vol. II A, W.P. Mason, Ed., Academic Press, New York, N.Y. 1965, pp. 281-349
11. R. Aronsson, H.E.G. Knape, and L.M. Torell, J. Chem. Phys. 68, 3794 (1978)
12. K.J. Rao, D.B. Helphrey and C.A. Angell, Phys. Chem. Glasses 14, 26 (1973)
13. A. Dietzel and H.P. Poegel, Proc. Intern. Glass. Congr. 3rd Venice, 319 (1963)

14. C.A. Angell, J. Phys. Chem. 68, 218 (1964)
15. F.S. Howell, R.A. Bose, P.B. Macedo and C.T. Moynihan, J. Phys. Chem. 78, 639 (1974)
16. C.A. Angell, in "Vibrational Spectroscopy in Molecular Liquids and Solids", Eds. E.M. Pick and S. Bratos, Plenum Press, 1980, p. 187
17. J.H.R. Clarke and S. Miller, Chem. Phys. Lett. 13, 97 (1972)
18. C.A. Angell and J. Wong, J. Chem. Phys. 51, 4519 (1969)
19. J. Wong, Ph.D. Thesis, Purdue University (1970)

THE KNIGHT SHIFT, RESISTIVITY AND THERMOCHEMICAL PROPERTIES OF COMPOUND FORMING MOLTEN ALLOYS DESCRIBED BY CHEMICAL THERMODYNAMICS

W. Geertsma and W. van der Lugt

Solid State Physics Laboratory, Materials Science Center
University of Groningen, Melkweg 1, 9718 EP Groningen
The Netherlands

In this paper we will discuss the Knight shift, resistivity and its respective temperature coefficients of a number of compound forming liquid alloy systems from the point of view of chemical equilibrium theory. The Knight shift calculations are based on the assumption of complete motional narrowing of the NMR resonance line. Matthiessen's rule will be used for the calculation of the resistivity. The various parameters entering the theory are chosen such as to fit the heat of mixing and volume of mixing; the fits are not optimized. It turns out that the Knight shift of Li-In can be described by one chemical equilibrium while for Li-Pb we need two chemical equilibria. The Flory model is used to describe the non-ideal behaviour of the liquid.

Model formulation

The most important results of chemical equilibrium theory for a system of n chemical equilibria in a A_cB_{1-c} system have already been given by Bhatia et al. (1). We will distinguish two types of chemical equilibria: the compound forming equilibrium $mA + nB \rightleftarrows A_mB_n$ and the ionic equilibrium $mA + nB \rightleftarrows mA' + nB'$. The first equilibrium takes into account charge transfer as well as chemical short range order, the second only charge transfer. The fraction of the various components are: for the compound forming equilibrium A: $a_1 = c - m\alpha$; B: $a_2 = 1 - c - n\alpha$; A_mB_n: $a_3 = \alpha$, and for the ionic equilibrium: A': $a_3 = m\alpha$, B': $a_4 = n\alpha$. The fraction α can be found as the solution of the chemical equilibrium equation in the Flory model:

$$L(c) = (a_1)^{m_i}(a_2)^{n_i}/(a_{i_1})^{m_{i_1}}(a_{i_2})^{n_{i_2}}$$

$$= K_{oi}(\varepsilon_{i_1})^{m_{i_1}}(\varepsilon_{i_2})^{n_{i_2}}/(\varepsilon_1)^{m_i}(\varepsilon_2)^{n_i}\exp(Q(c)), \qquad (1)$$

where: $K_{oi} = \exp(-g_i^{(o)}/RT)$;

$Q = m_i(r_1 + 1) + n_i(r_2 + 1) - (r_{i_1} + 1)m_{i_1} - (r_{i_2} + 1)n_{i_2}$;

$r_i = \sum_{j\neq i} a_j v_{ij}/RT$, v_{ij} and ε_i are Flory model parameters and L(c) is the equilibrium constant. The m_i's are the coefficients in the chemical equilibrium equations.

From the solution of these equations all thermodynamic properties can, in principle, be calculated. We neglect the temperature dependence of the Flory model parameters. The heat of mixing and volume of mixing become:

$$H_M = -RT \sum_i \alpha_i \frac{\partial \ln K_{oi}}{\partial T} + \sum_{i,j} a_i a_j v_{ij} \qquad (2^a)$$

$$V_M = - \sum_i \alpha_i (\partial g_i^{(o)}/\partial P)_{T,a_i} \qquad (2^b)$$

Our approach differs from the approach of Predel et al. (2) in that they neglect the first term in H_M and they decouple the calculation of H_M and the solution of the equilibrium equation.

When the liquid is described by one equilibrium all properties depend on only one equilibrium variable α, and so we can relate these properties, independent of L(c). Only the type of equilibrium and the composition of the compound remain as variables.

Assuming motional narrowing of the NMR resonance line, the Knight shift on nucleus A and B is given by

$$K_A = \frac{1}{c}\left[a_1 K_A^{(o)} + m\alpha K_A^{(1)}\right] \qquad (3^a)$$

$$K_B = \frac{1}{1-c}\left[a_2 K_B^{(0)} + n\alpha K_B^{(1)}\right] \qquad (3^b)$$

$K_i^{(o)}$ (i = A,B) is the Knight shift of pure i and $K_i^{(1)}$ is the Knight shift of the component i in the compound or ion.

The above equations can be rewritten as:

$$x_A/x_B = m(1 - c)/nc \qquad (4)$$

where $x_i = (K_i - K_i^{(o)})/(K_i^{(1)} - K_i^{(0)})$. When we measure one of the Knight shifts, the other one can be calculated when n, m and $K_i^{(1)}$, $K_i^{(0)}$ are fitted.

To calculate the resistivity we have used Matthiesen's rule (see also Hoshino and Young (3)). The resistivity is written as a sum of scattering events.

For the compound forming equilibrium we can write in this approximation

$$\rho = \frac{m}{e^2} \frac{V}{Z} \frac{1}{a}\left(\frac{a_1}{\tau_1} + \frac{a_2}{\tau_2} + \frac{a_3}{\tau_3}\right) \qquad (5)$$

where $a = \sum_i a_i = 1 - \alpha(m + n - 1)$, and Z is the number of free electrons per atom: $Z = a_1 Z_A + a_2 Z_B$, where Z_A and Z_B are the valencies of A and B respectively. τ_1, τ_2 and τ_3 are the relaxation times of electrons in A, B and the compound respectively. We have supposed that all electrons in the neutral molecule $A_m B_n$ can be considered as localized for the calculation of the resistivity. Once the Knight shift has been fitted, ρ can be calculated by fitting τ_1, τ_2 and τ_3 to the resistivity of the pure components and the maximum of the resistivity.

When the equilibrium constant approaches zero we can distinguish two cases. For $c > c_c = m/(n + m)$, c_c is the composition of the compound, we obtain for the Knight shift and resisitivity

$$K_A = K_A^{(0)} - \frac{m}{n}\left(\frac{1 - c}{c}\right)\left(K_A^{(0)} - K_A^{(1)}\right) \tag{6^a}$$

$$K_B = K_B^{(1)} \tag{6^b}$$

$$\rho = \frac{mV}{e^2}\left\{\left[n\left(c + (1 - c)\frac{1 - m}{n}\right)\tau_1\right]^{-1} + (1 - c)\left[n^2\left(c - \frac{m}{n}(1 - c)\right)\left[c + \frac{1 - c}{n}(1 - m)\right]\tau_3\right]^{-1}\right\} \tag{6^c}$$

For $c < c_c$ we find

$$K_A = K_A^{(1)} \tag{7^a}$$

$$K_B = K_B^{(0)} - \frac{n}{m}\left(\frac{c}{1 - c}\right)\left(K_B^{(0)} - K_B^{(1)}\right) \tag{7^b}$$

$$\rho = \frac{mV}{e^2}\left\{\left[m\left(1 - c + c\,\frac{1 - n}{m}\right)\tau_2\right]^{-1} + c\left[m^2\left(1 - c - \frac{n}{m}c\right)\left(1 - c + \frac{1 - n}{m}\,c\right)\tau_3\right]^{-1}\right\} \tag{7^c}$$

The most pronounced behaviour of the resistivity will come from the second term due to the vanishing of the number of free electrons Z. From the above equations one finds that in a plot of ρ as a function of K_A we can distinguish two branches, one for $c < c_c$, one for $c > c_c$.

This has also been found for the liquid systems Li-Pb, Li-Sn, Li-In. Such a plot is characteristic for systems described by one equilibrium. It is not in agreement with the rule $K^2 \propto \sigma$ given by Mott(4). In fig. 1 the Knight shift of A and B are plotted as a function of concentration. Clearly two concentration regions can be distinguished for small equilibrium constant L:

I) $c < c_c$, in which K_A is strongly dependent on composition and K_B

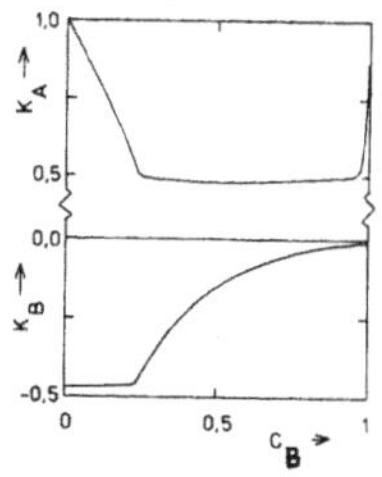

Fig. 1. Knight shift as a function of fraction B for an ideal solution with $L = .5 \times 10^{-6}$ with compound A_3B. $K_A^{(o)} = 1$; $K_B^{(o)} = 0$; $K_A^{(1)} = .5$; $K_B^{(1)} = -.48$.

is almost constant.

II) $c > c_c$ in which K_A is almost constant and K_B is strongly dependent on composition.

One sees immediately that this behaviour is predicted by the approximate formula $6^{a,b}$ and $7^{a,b}$. This is just what has been observed for the Knight shifts of Li and In in Li-In, of Li in Li-Ge, Li-Pb and Li-Sn and of Na in Na-Sn (5,6).

Finally we discuss the temperature dependences of the Knight shift and resistivity; neglecting the temperature dependence of the Knight shift of the pure components and compounds. We can derive:

$$\partial K_A/\partial T = \frac{m}{c}\frac{\partial \alpha}{\partial T}\left(K_A^{(1)} - K_A^{(o)}\right) \qquad (8^a)$$

$$\partial K_B/\partial T = \frac{nc}{m(1-c)}\frac{dK_A}{dT}\left(\frac{K_B^{(1)} - K_B^{(o)}}{K_A^{(1)} - K_A^{(o)}}\right) = \frac{nc}{m(1-c)}\frac{x_A}{x_B} \qquad (8^b)$$

Consequently when the temperature dependence of one of the Knight shifts is known the other one can be calculated. For the above model of compound forming liquids we expect $\partial K_A/\partial T > 0$ because $K_A^{(1)} < K_A^{(o)}$ and $\partial\alpha/\partial T < 0$.

For the temperature coefficient of the resistivity we derive ($\partial\tau_i/\partial T = 0$, $i = 1,2,3$):

$$\beta_R = \frac{1}{R}\frac{dR}{dT} = \beta_V - \beta_Z - \beta_a + \left\{\tau_c\left(\frac{c}{\tau_1} + \frac{1-c}{\tau_2} + \frac{\alpha}{\tau_c}\right)\right\}^{-1}\frac{\partial\alpha}{\partial T} \qquad (9)$$

with $\tau_c^{-1} = \left(\frac{1}{\tau_3} - \frac{m}{\tau_1} - \frac{n}{\tau_2}\right)$, where in general $\beta_Y = \frac{1}{Y}\frac{dY}{dT}$.

The volume expansion coefficient β_V is about 10^{-4} whereas $\beta_R \simeq 10^{-3}$ so β_V can be neglected. From eq. 9 we expect β_R to be negative around the compound forming concentration because β_Z as well as β_a are negative when $d\alpha/dT$ is negative and also the last term is negative when $1/\tau_3 > m/\tau_1 + n/\tau_2$.

The latter condition implies strong scattering by the compound relative to the pure A and B components. When the scattering by the compound is weak the last term in eq. 9 gives a positive contribution. The sign of β_R is determined by the change in the number of charge carriers and the change of the scattering strength.

We can now relate β_{K_A} and β_R. For $L \to \infty$ $(\alpha \to 0)$, $m = 4$, $n = 1$, $\tau_c \simeq \tau_1 \simeq \tau_2$, $K_A^{(o)}/(K_A^{(1)} - K_A^{(o)}) \simeq -2$, we obtain at the compound forming composition $c_c = {}^4/_5$:

$$\beta_{K_A} \simeq -{}^1/_5\,\beta_R \tag{10}$$

Note that this result is similar to the result obtained for the simple model $K^2 \propto \sigma$ so $\beta_K = -\frac{1}{2}\,\beta_R$. This result is in reasonable agreement with the observed temperature dependence of the resistivity and Knight shift of Li-Sn and Li-Pb.

In the other limit $K \to 0$ one finds that the temperature coefficient of the resistivity is much larger than that of the Knight shift.

In the limit $c \to 0$ we obtain for the Knight shift in the compound forming equilibrium for $m = 1$: $K_A \to \{K_A^{(o)} - \frac{1}{L'+1}(K_A^{(o)} - K_A^{(1)})\}$ where $L' = K_o(\varepsilon_3/\varepsilon_1^m\varepsilon_2^n)\exp[Q(c=0)]$. In the case $m \geq 2$ we obtain $K_A \to K_A^{(o)}$ for $c \to 0$. Similar expressions can be derived for the limit $c \to 1$. The essential thing is that $K_A \to K_A^{(o)}$ in both limits ($c \to 1$ and $c \to 0$). This is in disagreement with experiment, where K_A $(c \to 0)$ approaches about $\frac{1}{2}K_A^{(o)}$. This is one of the reasons we have also studied the ionic equilibrium system. This equilibrium gives the same equations and results for K_A and K_B except for the limiting behaviour $c \to 0$; in the ionic case one finds K_A $(c \to 0) \to K_A^{(1)}$. We expect the compound forming equilibrium to be valid around the compound forming composition and the ionic equilibrium, in which no chemical short range order is accounted for, to be valid in the other regions. Such a two equilibrium system will be discussed elsewhere.

Acknowledgements

This work forms part of the research program of the "Stichting voor Fundamenteel Onderzoek der Materie" (Foundation for Fundamental Research on Matter - FOM) and was made possible by financial support from the "Nederlandse Organisatie voor Zuiver Wetenschappelijk Onderzoek" (Netherlands Organization for the Advancement of Pure Research - ZWO).

References

(1) Bhatia, A.B., Hargrove, W.H., Phys. Rev. B10, (1974) 3186
(2) Predel, B. and Oehme, G., Z. Metallk., 81 (1979) 450.
(3) Hoshino, K., and Young, W.H., J. Phys. F, 10 (1980) L193.
(4) Mott, N.F. and Davis, E.A., "Electronic Processes in Non-crystalline Materials" 2nd ed. (Clarendon Press, Oxford, 1979)p. 196.
(5) Marel, C. van der, Oosten, A.B. van et al., J. Phys. F, to be published.
(6) Marel, C. van der, Oosten, A.B. van, et al., J. Phys. F, to be published.

ELECTRONIC AND STRUCTURAL PROPERTIES OF SOME COMPOUND-FORMING ALKALI-BASED ALLOYS

by B.P. Alblas, C. van der Marel*), W.Geertsma
and W.van der Lugt

Solid State Physics Laboratory, Materials Science Center,
University of Groningen, Melkweg 1, 9718 EP Groningen,
The Netherlands.

We have investigated experimentally several properties of the liquid binary systems Na-Pb, Na-Sn, Li-Ge and Li-Sn. For all of these systems strong evidence for the formation of compounds at approximately stoichiometric compositions has been found. The experiments include measurements of the Knight shift and the electrical resistivity as well as neutron diffraction.

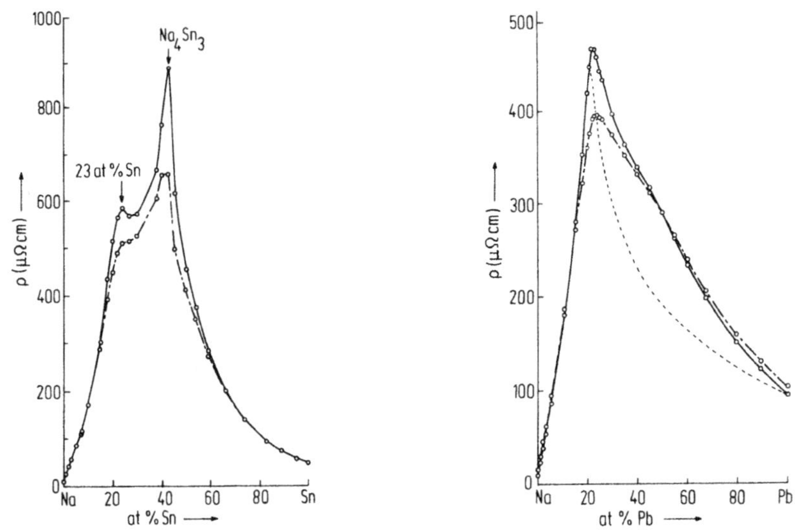

Fig. 1a,b. Resistivities of liquid Na-Sn and Na-Pb alloys, resp. at 10°C and 70°C, 10°C and 150°C above the liquidus.

*) *Present address:* Institut Laue-Langevin, B.P. 156X, F-38042 Grenoble Cedex, France.

As an example, in fig. 1 the resistivities, ρ, of the Na-Sn and Na-Pb systems are given. An analysis of the measurements shows that, at compositions close to the resistivity maxima, the mean free path of the conduction electrons is reduced to its minimum value, which is the atomic distance. According to Mott's theory, any variation of the resistivity corresponds then to a change in the density of states of the electrons at the Fermi level, $N(E_F)$. This is confirmed by Knight shift (K) measurements which exhibit a dip at such compositions, indicating a reduction of the density of states (Figs. 2 and 3). Previously, we have shown [1], that such dips are related to the resistivity maxima, as both originate from a reduction of $N(E_F)$.

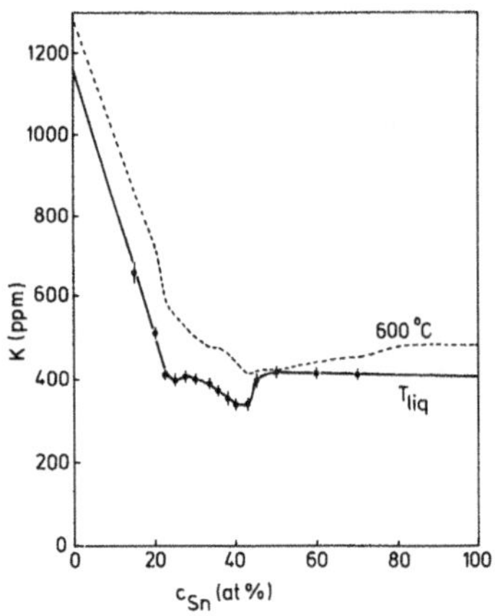

Fig. 2. Na Knight shift of liquid Na-Sn alloys _____ liquidus temperature .... isotherm, 600°C.

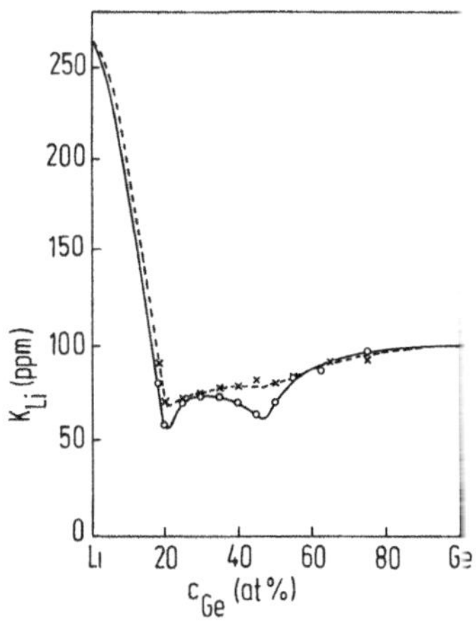

Fig. 3. Li Kright shifts of liquid Li-Ge alloys. _____ liquidus ----- 800°C.

The Knight shift is proportional to the spin susceptibility and, consequently, approximately proportional to $N(E_F)$. The conductivity, according to the Kubo-Greenwood formalism in the R.P.A. approximation, should be proportional to $N^2(E_F)$.

Most peculiar is the behaviour of the temperature dependence of the resistivity, $d\rho/dT$ (Fig. 4). It is negative and attains values of minus several $\mu\Omega$ cm per degree Kelvin. Apart from the sign, it is the absolute value of $d\rho/dT$ which is remarkable as it is at least two orders of magnitude larger than is usual in common metals or alloys. In a large composition interval of the alloys systems, ρ can be described by an exponential function $\rho = \rho_0\ e^{+E_0/kT}$. The interpretation of this seemingly thermal activated process is not straightforward, but it is plausible to assume that the negative sign of $d\rho/dT$ originates from an increase of $N(E_F)$ with increasing temperature. Once more, this is reflected in the Knight shift: dK/dT exhibits pronounced maxima at the "stoichiometric" compositions. (Fig. 5 and 6).

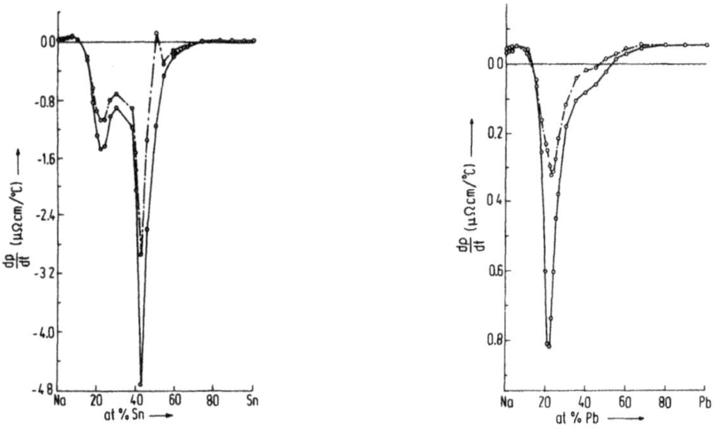

Fig. 4a,b. dρ/dT of Na-Sn and Na-Pb alloys for the same temperatures as in fig. 1.

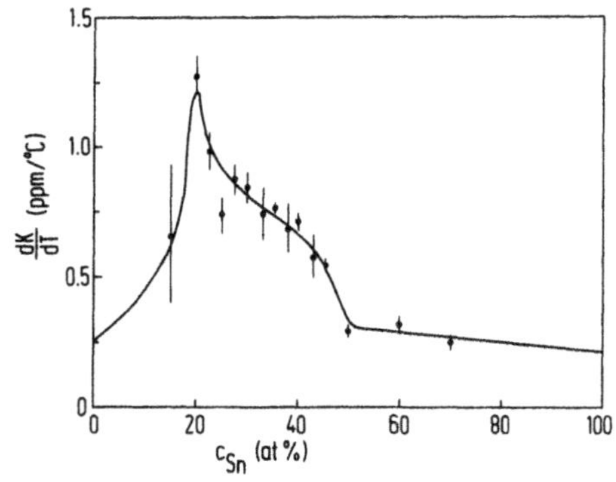

Fig. 5. Temperature dependence of the Knight shift, dK/dT, for Na-Sn alloys.

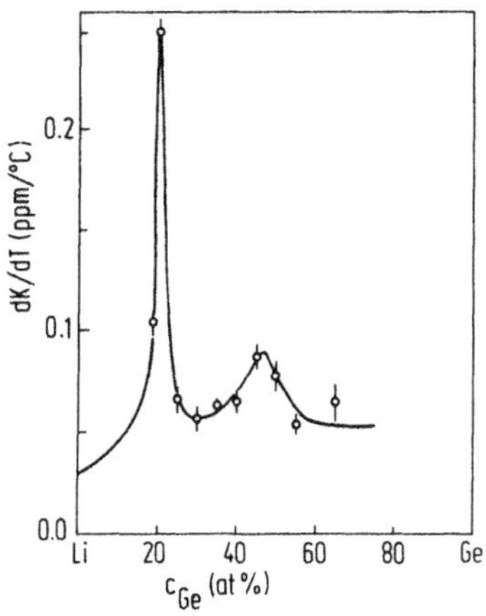

Fig. 6. Temperature dependence of the Knight shift, dK/dT, for Li-Ge alloys.

For liquid Na-Sn alloys, the resistivity and Knight shift data indicate strong compound formation at the approximate compositions Na_4Sn and Na_4Sn_3. The neutron diffraction results for Na_4Sn_3 exhibit a most pronounced prepeak (Fig. 7). It is interesting to compare this result with the crystal structure of the solid compound Na-Sn.

In this compound, covalent tetrahedra of Sn_4 occur, presumably negatively charged, with sodium ions adjacent to the faces of the tetrahedra. The distance between the centers of mass of the tetrahedra corresponds precisely to the prepeak in liquid Na_4Sn_3 suggesting the occurrence of a similar structure in the liquid. At other compositions (e.g. $NaSn_4$) a shoulder arises at the right-hand side of the main peak, which corresponds to Sn-Sn distances within a tetrahedron. But, unfortunately, when passing through the composition range, the shoulder appears as the prepeak disappears, so that the evidence for similarity of the liquid and the solid is not yet consistent. For liquid Li-Sn alloys, the resistivity and Knight shift results indicate a compound Li_4Sn. As Li_5Sn_2 is a "zero alloy" we could obtain the $\rho_{CC}(r)$ by neutron diffraction. Indeed, it shows the strong oscillations about zero which are characteristic for a high degree of order in the liquid (2), providing evidence for strong ionicity in liquid Li-Sn alloys (Fig. 8).

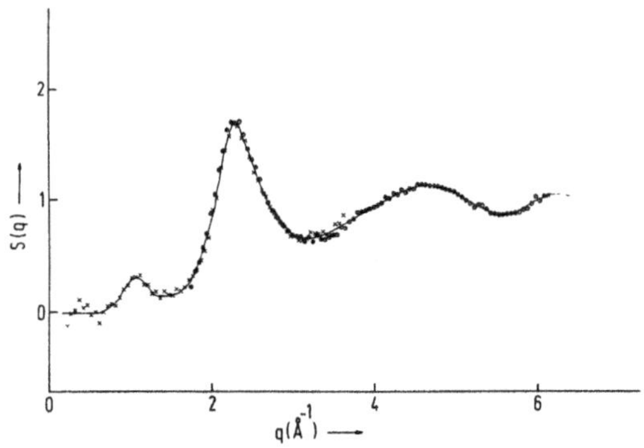

Fig. 7. Normalized Neutron Diffraction Intensity $S(q)$ For liquid Na_4Sn_3

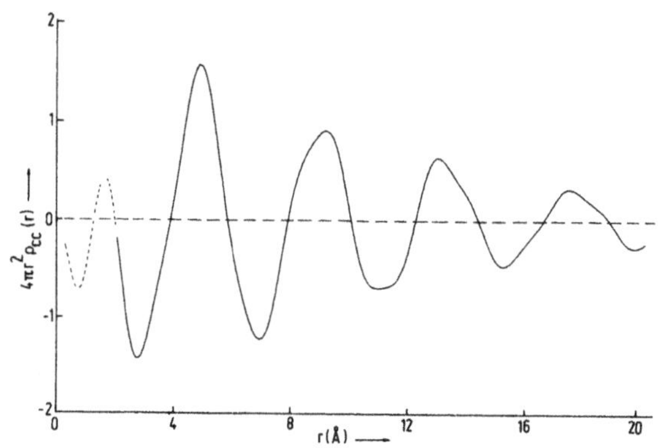

Fig. 8. Radial concentration-correlation Function $4\pi r^2 \rho_{cc}(r)$ of liquid Li_5Sn_2

Most of the properties of these alloys can be qualitatively understood by assuming the following simple band picture, which is intermediate between those for a liquid metal and a liquid salt.

If the temperature rises, the order in the liquid decreases, leading to a broadening of the bands corresponding to the Mott-Anderson picture. Also the contribution of the Madelung energy to the stability of the ionic liquid decreases. Consequently the gap fills up and N.F.E. metallic properties are gradually restored at higher temperatures. (Fig. 9)

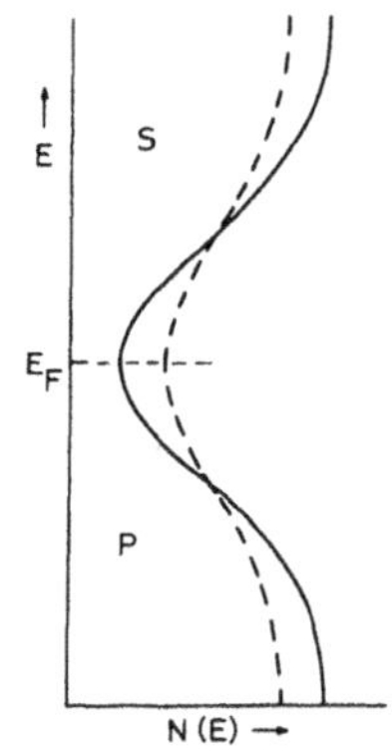

Fig. 9. Simple model for density of states.
—— lower temperature
---- higher temperature

We acknowledge with thanks the assistance during the experiments by G.J.Vinke, A.B.van Oosten, P.C.Stein and J.Dijkstra.
This work forms part of the research program of the "Stichting voor Fundamenteel Onderzoek der Materie" (Foundation for Fundamental Research on Matter-FOM) and was made possible by Financial support for the "Nederlandse Organisatie voor Zuiver Wetenschappelijk Onderzoek" (Netherlands Organization for the Advancement of Pure Research-ZWO).

References

[1] C. van der Marel, W. Geertsma and W. van der Lugt, J. Phys. F 10 (1980) 2305.
[2] P. Chieux and H. Ruppersberg, J. de Phys. C8, 41 (1980) 145.

A SIMPLE MODEL FOR COMPOUND FORMING LIQUID ALLOY SYSTEMS

W. Geertsma

Solid State Physics Laboratory, Materials Science Center
University of Groningen, Melkweg 1, 9718 EP Groningen,
The Netherlands

We discuss the resistivity, thermopower and Knight shift as a function of temperature and composition for a number of alkali-IV^A metal alloys using the tight binding C.P.A. model. Elastic electron- phonon interaction results in a strong temperature dependence of the electronic structure. The calculations support the experimental evidence of near localization of the electrons close to stoichiometric compositions in liquid LiPb, LiSn, NaSn, NaPb and LiGe.

Recently the Knight shift and electronic transport properties of a number of liquid alkali metal-IV^A metal alloys have been measured as a function of temperature and concentration (1,2). (These liquid alloys exhibit partial salt-like character.) The properties of these alloys show anomalous behaviour around the composition of compound formation: a sharp maximum in the resistivity and a sharp negative minimum in its temperature derivative, a distinct minimum in the Knight shift and a maximum in its temperature derivative. The thermopower changes sign at this concentration.

We will show that these properties can be explained within a simple tight binding band model in the coherent potential approximation (C.P.A.). Short range chemical order is neglected. By including elastic electron-phonon interaction we can simulate effects due to (increasing) thermal disorder in the liquid. Increasing thermal disorder leads to a broadening of the bands and a filling of the (pseudo) gap.

Theory, model and calculation procedure

We have used the C.P.A. formalism as derived by Chen, Weisz and Sher (3). The basic equation to be solved is:

$$F(z) = \int d\eta \, \{\frac{x}{\sqrt{2\pi\alpha_A}} e^{-\eta^2/2\alpha_A} (z - \varepsilon_A - \eta - \tfrac{1}{4}F)^{-1}$$

$$+ \frac{y}{\sqrt{2\pi\alpha_B}} e^{-\eta^2/2\alpha_B} (z - \varepsilon_B - \eta - \tfrac{1}{4}F)^{-1}\} \qquad (1)$$

with ε_i $(i = A,B)$ the random atomic potentials; $x = 1 - y$ is the fraction of A, α_i (i = A,B) are the electron-phonon interaction parameters, which we assume to be proportional to the temperature (3).

The self energy is given by

$$\Sigma = z - 1/F - \tfrac{1}{4}F \tag{2}$$

for a semicircular unperturbed band. The integral equation 1 was solved by iteration using the result for $\alpha_A = \alpha_B = 0$ as input and then increasing α_A, α_B stepwise to the desired value.

The density of states is given by

$$N_i(\varepsilon) = -\frac{1}{\pi}\int d\eta \frac{x}{\sqrt{2\pi\alpha_i}} I_m F_i(\eta, \varepsilon + i\delta) e^{-\eta^2/2\alpha_A}, \tag{3}$$

with: $F_i(\eta, z) = (z - \varepsilon_i - \eta - \tfrac{1}{4}F)^{-1}$.

The total density of states becomes $N(\varepsilon) = xN_A(\varepsilon) + yN_B(\varepsilon)$. The charge transfer is given by:

$$\Delta Q_i = Z_i - 2\int d\varepsilon f(\varepsilon) N_i(\varepsilon) \equiv Z_i - Q_i \tag{4}$$

We approximate the Knight shift by

$$K_i = Q_i \times N(\varepsilon_F) \tag{5}$$

where ε_F is the Fermi level.

The conductivity is given by:

$$\sigma = \sigma_o \int d\eta(-\partial f(\eta)/\partial\eta)\frac{N^3(\eta)}{\Delta(\eta)}\left(1 + \frac{6}{\pi}\frac{\Delta(\eta)}{N(\eta)}\right) \equiv \int \sigma(\eta)\left(-\frac{\partial f(\eta)}{\partial\eta}\right)d\eta \tag{6}$$

with $\Delta(\eta) = |I_m\Sigma(\eta + i\delta)|$ and $\sigma_o = (e^2hv_m^2\pi^2)/12\Omega$, v_m is the maximum velocity in the band, Ω is the atomic volume. The thermopower is calculated by:

$$S = \frac{k_B}{e\sigma}\int \sigma(\eta)\left(\frac{\eta - \varepsilon_F}{k_BT}\right)(-\partial f(\eta)/\partial\eta)d\eta = \frac{\pi^2k_B^2T}{3}\left.\frac{d\ \ln\sigma(\eta)}{d\eta}\right|_{\eta=\varepsilon_F} \tag{7}$$

The latter equality holds at low temperatures.

For each concentration we have performed a self consistent calculation. In this respect, it differs from the calculations performed by Chen et al. (3) and Richter and Schiller (4). The first authors only performed calculations for x = .1 and .5. The latter used a Lorentzian distribution function for the fluctuation of the atomic potentials while we use a Gaussian.

Results and discussion

In figs. 1, 2, 3, 4 and 5 we present some of the results of our calculations for $\varepsilon_A = -\varepsilon_B = .4$ for various values of $\alpha = \alpha_A = \alpha_B$. More details will be given in a forthcoming paper. For this value of ε_A the bands just overlap for x = .5. The resistivity increases with a factor 10 for $\alpha = .015$ from x = 0 to x = .5. This is comparable to the results obtained for the liquid alkali-IV^A metal alloys mentioned in the introduction.

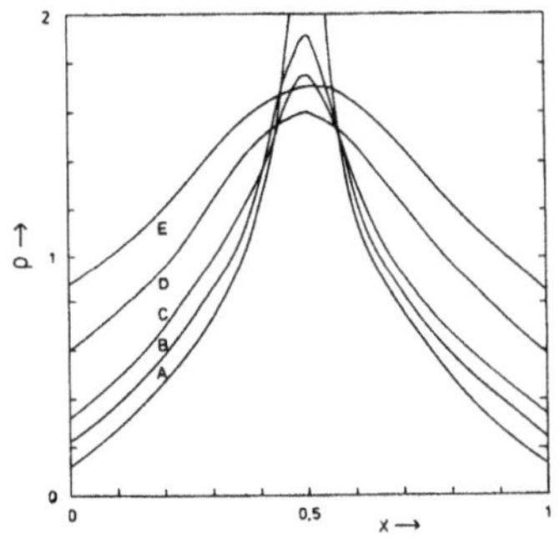

Fig. 1. Resistivity as a function of concentration for $\varepsilon_A = .4$ for $\alpha = \alpha_A = \alpha_B$ is A: .007, B: .015, C: .031, D: .049, E: .099. ρ is in units ρ_o.

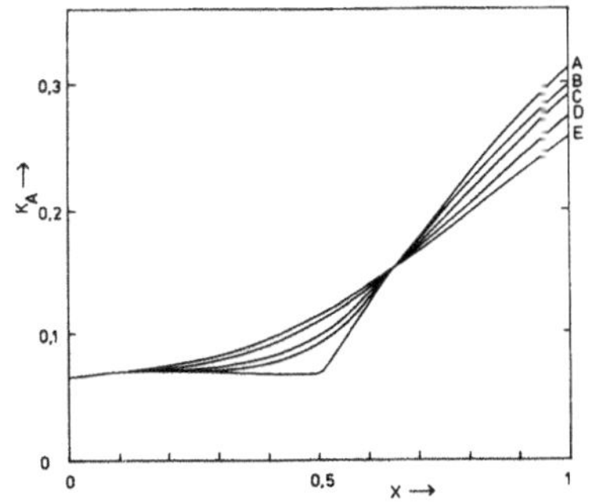

Fig. 2. Knight shift as a function of concentration for $\varepsilon_A = .4$ for $\alpha = \alpha_A = \alpha_B$ is A: 0.0, B: .031, C: .049, D: .099, E: .159.

In fig. 4 we have plotted the temperature coefficient of the resistivity $\beta_R = \frac{1}{\rho}\, d\rho/dT$. Around the compound forming concentration we find a negative temperature coefficient, which itself depends on the temperature. This strong temperature dependence of the resistivity is due to

i) a small increase of the density of states in the gap region with α (temperature),

ii) a strong decrease of the scattering life time ($\sim \Delta^{-1}$) of the electrons with increasing α (= temperature).

The calculated temperature coefficient of the resistivity is in the gap region of the same order as the experimental values for Li-Pb, Li-Sn, Na-Pb and Na-Sn. Also the experimentally found decrease of β_R with increasing temperature is found in our model. The thermopower is plotted in fig. 3. The thermopower changes sign at x = .5. The maximum (minimum) of the thermopower shifts away from x = .5 with increasing α.

At low temperatures the thermopower increases with temperature

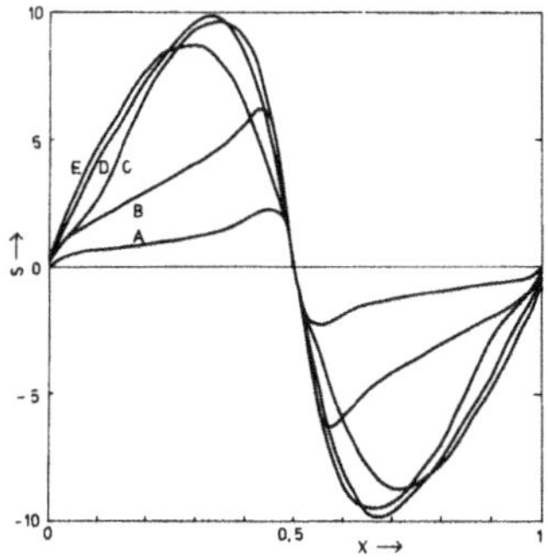

Fig. 3. Thermopower (S') as a function of concentration for $\varepsilon_A = .4$ for $\alpha = \alpha_A = \alpha_B$ is A: 0.038, B: .015, C: .060, D: .090, E: .184.

whereas at high temperatures it decreases with temperature.

We have also compared the results of the two equations for the thermopower. We found that at high temperatures ($\alpha \gtrsim .015$) the approximate formula for the thermopower gives a result which is about 30-50% too high.

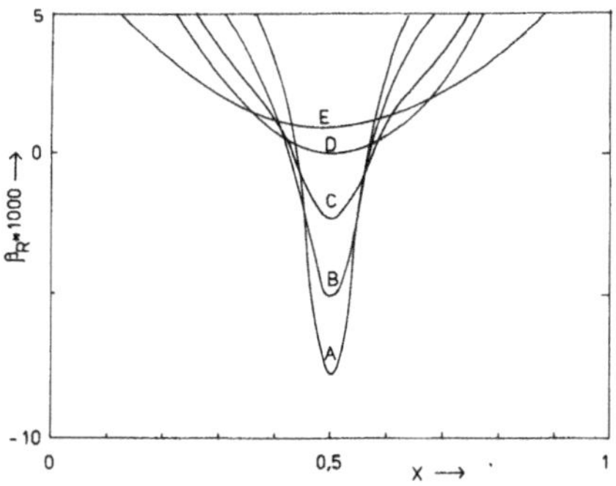

Fig. 4. Temperature coefficient of the resistivity β_R for $\varepsilon_A = .4$ for $\alpha = \alpha_A = \alpha_B$ is A: .015, B: .031, C: .049, D: .099, E: .159.

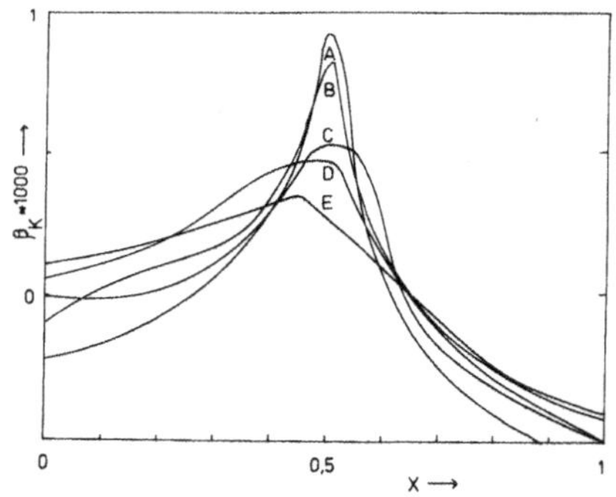

Fig. 5. Temperature coefficient of the Knight shift β_K for $\varepsilon_A = .4$ for $\alpha = \alpha_A = \alpha_B$ is A: .015, B: .031, C: .049, D: .099, E: .159.

The calculated Knight shift K_A is plotted in fig. 2. For small α, K_A depends linearly on the concentration up to $x \simeq .5$. For $x > .5$ the Knight shift increases strongly with increasing x. Around the compound forming composition x = .5 the Knight shift increases strongly with increasing temperature (fig. 5). $\beta_K = (1/K_A)(dK_A/dT)$, the temperature coefficient of the Knight shift, has a maximum at x = .5 of the order of 10^{-3}, which is of the order of magnitude found for the

Li Knight shift in Li-Pb, Li-Sn (1,2).

The linear dependence of K_A on x for x < .5 is due to the cancelling of the increase of Q_A by the decrease in $N(\varepsilon_F)$ with increasing x. The strong increase of K_A for x > .5 is due to the increase of $N(\varepsilon_F)$ and Q_A with increasing x. The positive β_K is due to Q_A as well as to $N(\varepsilon_F)$. For $\delta \equiv K_A(x=0)/K_A(x=1)$ we obtain a value of about .25 while experimentally one obtains for the systems Li-Pb, Li-Sn, Li-Ge and Na-Sn about .4 to .5.

A slight increase of ε_A from .4 to .5 is sufficient to increase β_K (x= .5) with a factor 2 while β_R remains nearly the same. Also the maximum in the resistivity increases strongly; for $\alpha = 0$ the system is an insulator. The Knight shift shows a minimum at x = .5 for ε_A = .5 at low temperatures (low α) because of the appearance of a very low d.o.s. for small values of α.

From the above discussion of our results we conclude that a simple tight binding model can describe well the qualitative features of alkali-IV^A metal alloys in the gap region. The temperature dependence of the calculated quantities cannot be explained in this model for x > .8 and x < .2, because we only consider one band.

We have neglected chemical short range order (c.s.r.o.) and this implies neglecting the concentration dependence of ε_A. In the region of the compound forming concentration ε_A will become larger with increasing c.s.r.o. This will give an extra contribution to the temperature coefficients: $|\beta_R|$ and β_K will increase while the peak in the resistivity will become larger and the dip in the Knight shift will become deeper.

Finally we conclude that the electrons in these disordered alkali-IV^A alloys are on the threshold of becoming localized; it should be emphasized, however, that the model has been chosen such as to exclude actual localization due to disorder.

Acknowledgements

This work forms part of the research program of the "Stichting voor Fundamenteel Onderzoek der Materie" (Foundation for Fundamental Research on Matter - FOM) and was made possible by financial support from the "Nederlandse Organisatie voor Zuiver Wetenschappelijk Onderzoek" (Netherlands Organization for the Advancement of Pure Research - ZWO).

REFERENCES

(1) C. van der Marel, A.B. van Oosten, W. Geertsma and W. van der Lugt,

J. Phys. F., to be published.
(2) C. van der Marel, A.B. van Oosten, W. Geertsma and W. van der Lugt, J. Phys. F., to be published.
(3) An-Ban Chen, G. Weisz and A. Sher, Phys. Rev. B5 (1972) 2897.
(4) J. Richter and W. Schiller, Phys. Stat. Sol. (b) 92 (1979) 511.

^{8}Li SPIN-LATTICE RELAXATION IN THE LIQUID ALLOYS Li-Bi and Li-Pb

G. Kiese, P. Heitjans, H. Ackermann, B. Bader, W. Buttler, P. Freiländer, C. van der Marel, H. Ruppersberg*, H.-J. Stöckmann

Fachbereich Physik, Univ. Marburg, W.-Germany, and Institut Laue-Langevin, Grenoble, France
*Fachbereich Angewandte Physik, Univ. Saarbrücken, W.-Germany

Abstract: The nuclear spin-lattice relaxation rate T_1^{-1} of β-active ^{8}Li ($T_{1/2}$ = 0.8 s) has been measured in liquid Li-Bi and Li-Pb alloys. T_1^{-1} as function of concentration and temperature shows marked deviations from metallic behaviour near the compositions Li_3Bi and Li_4Pb. Combining the relaxation data with Knight shift and/or conductivity data the mean 'residence time' τ_e of the conduction electrons at a nuclear site and the density of states at the Fermi surface $N(E_F)$ are obtained relative to the free-electron values. The results show that in liquid Li-Bi the tendency towards electron localization and compound formation is significantly stronger than in Li-Pb. The melt $Li_{0.8}Pb_{0.2}$ seems to become metallic above 1400 K.

1. INTRODUCTION

In the last few years an increasing number of experiments has been performed on binary liquid lithium-based alloys. This interest is caused to some extent by the observation of deviations from metallic behaviour around specific stoichiometric compositions in some of these systems.

The most extensively studied Li alloys belonging to this class are Li-Pb and Li-Bi which show anomalies in a number of properties near the compositions Li_4Pb and Li_3Bi. These anomalies are ascribed to compound forming tendencies in the melts. Properties already investigated in the two systems include electrical conductivity, density, and thermodynamic properties /1/. In the case of Li-Pb additionally neutron scattering /2,3/ as well as nuclear spin-lattice relaxation /4/ and Knight shift /5/ have been studied. These methods are capable to investigate the melt on a microscopic scale.

The present spin-relaxation study continues our work on Li-Pb /4/ and extends it to the system Li-Bi. The starting point for the discussion of the relaxation time T_1 is the well-known Korringa relaxation /6/

$$T_1 T K^2 = S_o \qquad (1)$$

where K is the Knight shift, T the temperature and $S_o = 1.20 \cdot 10^{-5}$ sK the Korringa constant of the present probe nucleus ^{8}Li (cf. /4/). In this paper we are particularly interested in deviations from this law expressed by the enhancement

η of the experimental relaxation rate over the Korringa rate. The latter one is calculated from eq. (1) using the experimental Knight shift:

$$\eta = \frac{(T_1^{-1})_{exp}}{(T_1^{-1})_{Korr}} = \frac{S_o}{(T_1 T K^2)_{exp}} \qquad (2)$$

As pointed out by Warren /7/ an enhancement $\eta > 1$ yields information on the increase of the mean 'residence time' τ_e of the conduction electrons on a nuclear site and the reduction of the density of states $N(E_F)$ at the Fermi surface as compared to the free-electron values. One can show /7/ that the approximate relation

$$\eta \simeq \frac{\tau_e}{\hbar\, N(E_F)} \qquad (3)$$

should hold.

An analysis of the T_1 data along these lines is possible in Li-Pb over nearly the whole concentration range and is done here also at temperatures above 1000 K thus extending our previous work. In Li-Bi the discussion of our T_1 data is hampered by the fact that Knight shift data are lacking arround the composition $Li_{0.75}\,Bi_{0.25}$ and are available only for small and high Bi concentrations.

2. MEASUREMENTS AND RESULTS

The experiments were carried out at the High-Flux Reactor Grenoble. Polarized β-active ^8Li nuclei were produced in the Li alloys by capture of polarized thermal neutrons. The ^8Li polarization and its decay constant $1/T_1$ after a neutron activation pulse was measured via the asymmetry of the ^8Li β-radiation distribution. (For details see /8/).

The Li-Pb and Li-Bi alloys were enclosed in stainless steel and niobium containers, respectively. One of the Li-Pb samples, viz. $Li_{0.8}\,Pb_{0.2}$, was prepared also with Nb as container in order to allow measurements on this alloy at temperatures substantially higher than 1000 K. The results obtained with the two $Li_{0.8}\,Pb_{0.2}$ samples at 1000 K showed no significant influence of the container material.

The concentrations of the investigated alloys ranged from 0 to about 80at% Pb and Bi, respectively. The temperature was varied from the respective liquidus temperature up to about 1350 K in Li-Pb and 1500 K in Li-Bi. Values of the applied magnetic field B_o were 30 mT and 334 mT. T_1^{-1} was found to be independent of B_o.

The positive slope of T_1^{-1} vs. temperature found at all compositions indicates that magnetic hyperfine interaction with conduction electrons is the dominant relaxation mechanism. Diffusion-induced dipolar and quadrupolar relaxation, which can be as important as relaxation by conduction electrons in solid Li alloys /9/, is expected to be negligible in the liquid alloys due to extreme motional averaging. The diffusion coefficient increases by a factor of more than hundred /10/ at the transition from the solid to the liquid state.

The reduced relaxation rate $(T_1T)^{-1}$ which should be essentially independent of temperature according to (1) exhibited a small but significant temperature dependence being fairly linear in T. In Fig. 1 the temperature derivatives $d(T_1T)^{-1}/dT$ are plotted vs. Bi-concentration in Li-Bi. Clearly, measurements are still lacking near 25at% Bi; but the data already suggest a maximum at about this concentration.

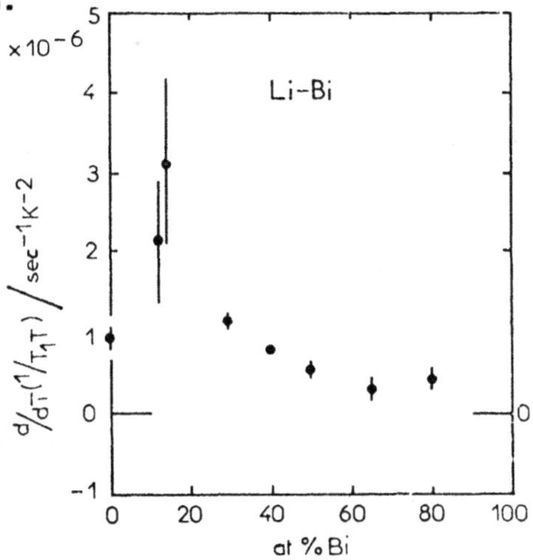

Fig. 1. Temperature derivative of the reduced spin-lattice relaxation rate $(T_1T)^{-1}$ of ^8Li versus concentration in liquid Li-Bi.

Fig. 2 illustrates the isothermal concentration dependencies of $(T_1T)^{-1/2}$ in Li-Pb and Li-Bi, respectively. The minima occur near the compositions $Li_{0.8}Pb_{0.2}$ and $Li_{0.75}Bi_{0.25}$, respectively. Despite the similar overall behaviour there is an interesting

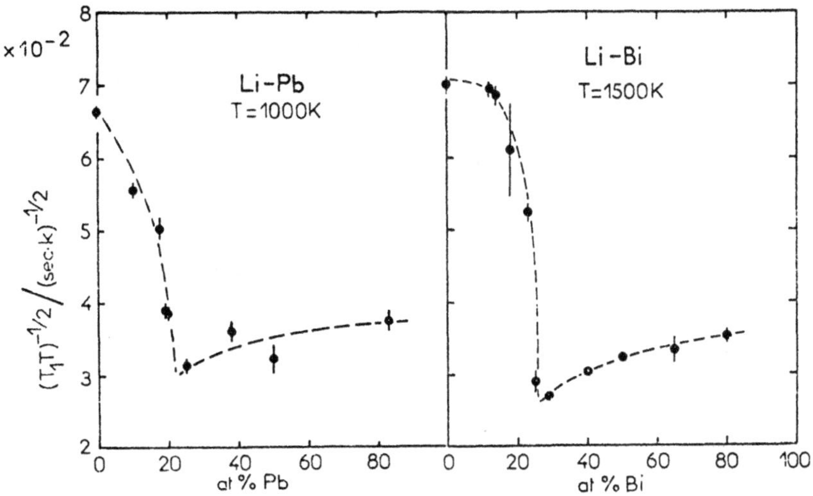

Fig. 2 Concentration dependence of $(T_1T)^{-1/2}$ of ^8Li in liquid Li-Pb and Li-Bi.

difference between the two systems: $(T_1T)^{-1/2}$ rapidly decreases upon addition of Pb to pure liquid Li, but remains approximately constant when Bi is added up to about 10at%. When the Bi-concentration exceeds this value, $(T_1T)^{-1/2}$ decreases rapidly to its minimum value.

3. DISCUSSION

Combining our T_1 data in Li-Pb with available Knight shift data /5/ the enhancement η according to (2) is obtained as a function of lead concentration c (Fig. 3).

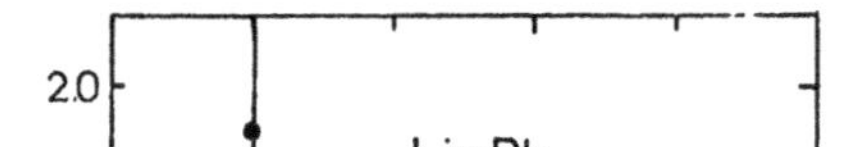

Fig. 3. Enhancement of ^{8}Li relaxation rate versus concentration in liquid Li-Pb

In the free-electron model one expects $\eta(c) \simeq 1$. Electron-electron interactions /11/ cause a reduction of η to a value <1. This is observed e.g. in liquid Li-Na where $\eta(c) \simeq 0.63$ at T = 673 K independent of concentration /4/. In liquid Li-Pb, however, η is found to rise upon addition of Pb to Li up to 1.94(24) at $c \simeq 20$at% Pb. The electrical resistivity, too, exhibits a pronounced maximum at about this composition /12/. The increasing Pb-concentration leads to a rapid decrease of the mean free path λ of the conduction electrons, until λ is about equal to the interatomic distance a. According to Mott /13/ the conductivity in this strong-scattering regime is related to the reduced density of states at the Fermi energy $g = N(E_F)/N(E_F)^{free}$ by

$$\sigma = \sigma_o^M g^2 \tag{4}$$

where $\sigma_o^M = e^2k_F^2a/3\pi^2\hbar$. On the other hand, following Warren /7/, σ in this regime is related to the enhancement η by

$$\sigma\eta = \sigma_o^W \simeq \frac{e^2a^2}{3\Omega\hbar} \tag{5}$$

(Ω = atomic volume).

With $\sigma_o^M \simeq \sigma_o^W$ we get

$$g \simeq \eta^{-1/2} \qquad (6)$$

Thus from the value η= 1.94 at c ≃20at% Pb and T ≃1000 K we can estimate g≃ 0.72. Using the experimental conductivity /12/ and density /14/, g = 0.75 at about 1000 K can be calculated with eq. (4) (see e.g. /15/). Using the measured total magnetic susceptibility and assuming ionic structure for Li_4Pb Hackstein /16/ estimated g = 0.60 at T ≃ 1170 K. This value appears to differ somewhat from our value considering the temperature dependence of η reported below.
The mean residence time τ_e of strongly scattered electrons at a nuclear site relative to that of free electrons, τ_e^{free}, can be obtained from eq. (3) and the approximate relation $\tau_e^{free} \simeq \hbar\, N(E_F)^{free}$. From this together with eq. (6) we get

$$\frac{\tau_e}{\tau_e^{free}} \simeq \eta^{1/2} \qquad (7)$$

which yields $\tau_e \simeq 1.4\ \tau_e^{free}$ for $Li_{0.8}Pb_{0.2}$ at 1000 K.

Furthermore, using the measured T-dependencies of $(T_1T)^{-1}$ and K /5/ $\eta(T)$ can be calculated. The result for $Li_{0.8}Pb_{0.2}$ is shown in Fig. 4. η decreases with increasing T, i.e. the compound forming tendency weakens at higher temperatures. Above 1400 K the melt seems to be metallic.

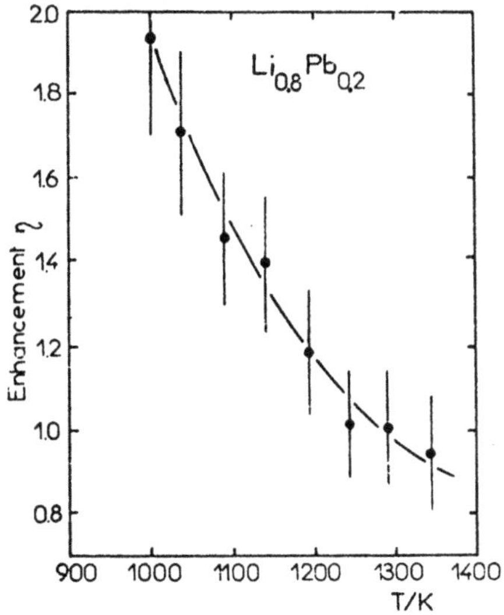

Fig. 4. Temperature dependence of the ^8Li relaxation rate enhancement in liquid $Li_{0.8}Pb_{0.2}$.

In liquid Li-Bi the enhancement η near the stoichiometric composition Li_3Bi can only be estimated since Knight shift data in the appropriate concentration range are lacking. Using eq. (5) we obtain from the measured conductivity and density /17/ for c ≃ 25at% Bi and T = 1500 K: η = 4.9. With this value for η and the measured T_1 we can make a crude estimation of the Knight shift in $Li_{0.75}Bi_{0.25}$ at T = 1500 K using eq. (2). This yields K ≃ 40 ppm near the expected minimum.

We further obtain from eqs. (6) and (7): g = 0.45 and $\tau_e = 2.2\ \tau_e^{free}$. The result for g in this case, unlike the one in $Li_{0.8}Pb_{0.2}$, agrees with the value obtained by Hackstein /16/ from the susceptibility under the assumption of ionic bonding (g = 0.47 at T = 1570 K).

Knight shift data in liquid Li-Bi are available for concentrations up to 5at% and beyond 50at% Bi /van der Marel, private communication/. Combining these data with our T_1 results, we get $\eta \simeq 0.7$ for the low and $\eta \simeq 1.0$ for the high Bi concentrations. The values are compatible with the free-electron model, the usual correction for electron-electron interactions included.

Concerning the temperature dependence of $(T_1T)^{-1}$, the increase of $d(T_1T)^{-1}/dT$ around the stoichiometric composition (s. Fig. 1) is not unexpected in view of the direct relation of $(T_1T)^{-1}$ to σ and K (cf. eqs. (2) and (5)). $d\sigma/dT$, being known for Li-Pb /12/ and Li-Bi/17/, and dK/dT, which is available for Li-Pb /5/, also exhibit marked peaks near the stoichiometric compositions.

The formerly mentioned difference in the behaviour of $(T_1T)^{-1/2}$ (c) at low concentrations of Pb and Bi (s. Fig. 2) may tentatively be explained by the assumption that in dilute Li-Pb alloys the 6p band of Pb is not completely filled which gives rise to a strong perturbation of the Li density of states. This is supported by the relatively large residual resistivity $\delta\rho$ of Li-rich liquid Li-Pb (see e.g. /15/). On the other hand one may conclude that in dilute Li-Bi alloys the Bi 6p band is completeley filled below E_F therefore not changing $N(E_F)$ of Li. In this context we refer to Avci and Flynn /18/ who ascribed the much smaller $\delta\rho$ of CsSb compared to CsSn alloys to complete filling of the outer shell of Sb.

Summarizing we may conclude that in liquid Li-Bi the tendency towards electron localization and compound formation is significantly stronger than in Li-Pb. This is also consistent with measurements of e.g. the enthalpy of mixing in the two systems /19,20/. The present results fit satisfactorily in the plot $\eta(\sigma)$ of Warren /7/ shown in Fig. 5.

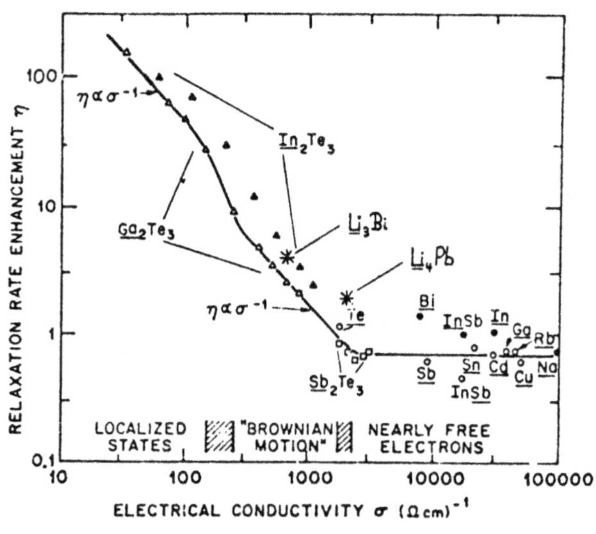

Fig. 5. Relaxation rate enhancement of the underlined atomic species versus conductivity (from Warren /7/). The presently determined η values in $Li_{0.8}Pb_{0.2}$ at 1000 K and in $Li_{0.75}Bi_{0.25}$ at 1500 K have been inserted.

Whereas $Li_{0.8}Pb_{0.2}$ is a borderline case between the nearly-free-electron and the strong-scattering regime, $Li_{0.75}Bi_{0.25}$ clearly falls into the latter region.

ACKNOWLEDGEMENT

We thank E. Jäger, D. Röhrich and A. Schirmer for help in preparing some of the samples. This work is sponsored by the Bundesministerium für Forschung und Technologie.

REFERENCES

/1/ For reviews see: F. Hensel, Adv. Phys. 28, 555 (1979); W.W. Warren, in Advances in Molten Salt Chemistry, Vol. 4, edited by J. Braunstein, G. Mamantov, and G.P. Smith (New York, Plenum Press 1980)

/2/ H. Ruppersberg, H. Egger, J.Chem.Phys. 63, 4095 (1975)

/3/ M. Soltwisch, D. Quitmann, H. Ruppersberg, J.B. Suck, Phys.Lett. 86A, 241 (1981)

/4/ P. Heitjans, G. Kiese, H. Ackermann, B. Bader, W. Buttler, K. Dörr, F. Fujara, H. Grupp, A. Körblein, H.-J. Stöckmann, J. Physique 41, C8-409, (1980)

/5/ C. van der Marel, W. Geertsma, W. van der Lugt, J. Phys. F10, 2305 (1980)

/6/ J. Korringa, Physica 16, 601 (1950)

/7/ W.W. Warren, Phys.Rev. B3, 3708 (1971)

/8/ H. Ackermann, Hyperfine Interactions 4, 645 (1978)

/9/ P. Heitjans, A. Körblein, H. Ackermann, D. Dubbers, F. Fujara, M. Grupp, H.-J. Stöckmann, Proc. 19th Congr. Ampere, Heidelberg, 1976 p. 281

/10/ N.H. Nachtrieb, Ber. Bunsenges.Phys.Chem. 80, 678 (1976)

/11/ A. Narath, H.T. Weaver, Phys.Rev. 175, 373 (1968)

/12/ V.T. Nguyen, J.E. Enderby, Phil.Mag. 35, 1013 (1977)

/13/ See e.g.: N.F. Mott, E.A. Davis, Electronic processes in noncrystalline materials (Oxford, Clarendon Press 1979)

/14/ H. Ruppersberg, W. Speicher, Z.Naturf. 31a, 47 (1976)

/15/ C. van der Marel, Thesis, Groningen 1981

/16/ K. Hackstein, Thesis, München 1980

/17/ G. Steinleitner, W. Freyland, F. Hensel, Ber. Bunsenges.Phys.Chem. 79, 1186 (1976)

/18/ R. Avci, C.P. Flynn, Phys.Rev. B19, 5967 (1979)

/19/ B. Predel, G. Oehme, Z. Metallkde 70, 450 (1979)

/20/ B. Predel, G. Oehme, Z. Metallkde 70, 618 (1979)

Correlation Effects in Metal Ammonia Systems

Juana V. Acrivos, San Jose State University, CA 95192
A. S. Robertson and M. P. Klein, Lawrence Berkeley Laboratory, U.C.B., CA 94720

Abstract

The recent structural and electrochemical measurements of the alkali metal-ammonia systems are reconciled in a model where the correlation energy leads to the formation of bound states. This is evident in the X-ray absorption of $Rb_x(NH_3)_{1-x}$ near the Rb K-edge as a white peak which appears only when the metal mole fraction is reduced below $x_c = 0.04$. These transitions must be associated with the correlation effects which lead to the formation of sodium anions predicted by electrochemical measurements. If the anions act as impurity centers with a minus two valence change from the uniformly charged positive background, Friedel oscillations in the electron density must be taken into account in a statistical description of the system. The deviations of the activity coefficients from the usual Debye-Hückel expression are discussed.

1. Introduction

Alkali metals dissolved in ammonia exhibit a variety of interesting electromagnetic phenomena as a function of the total metal mole fraction, x. As x increases, the paramagnetic infinitely dilute solutions ($x < 10^{-6}$) become diamagnetic near $x = 10^{-2}$ to 10^{-3} and finally display Pauli spin susceptibility for $x > x_c = 0.04$, where x_c is the critical mole fraction for a metal to insulator (MI) transition of the Mott type [1]. The purpose of this work is to reconcile the recent structural [2, 3] and thermodynamic data [4] and to describe the system above concentrations where the extended Debye-Hückel theory breaks down (near $x = 10^{-3}$). Section 2 presents the physical significance of the various semiempirical models used to describe the system. Section 3 shows that in metal ammonia solutions (MAS) the Boltzmann statistics used in the Debye-Hückel treatment of activity coefficients must be replaced by Thomas-Fermi statistics when $x > x_c$ and that when $10^{-3} < x < x_c$ the thermodynamic potential must include Friedel oscillations produed by the metal anions.

2. Structure and Thermodynamic Relations

The X-ray absorption spectra of $Rb_x(NH_3)_{1-x}$ [2] and the neutron scattering produced by $Na_x(ND_3)_{1-x}$ and $Li_x(ND_3)_{1-x}$ [3] near x_c indicate that the positively charged cations contribute as a uniformly charged background. However, the nature of localized and/or extended electron states can be determined from the absorption spectra of the cation, specially near the Rb K-edge. Figure 1 shows the absorption A and the first A' and second A" derivatives near 15 keV obtained by procedures described previously [2]. The white peak labeled A_1 has been assigned to final exciton states that appear at concentrations near x_c, because it is not observed in the metallic regime. The physical significance here is that the final states of the transition from an initial cation 1s state are influenced by the solvated electron states which must contribute to the nuclear shielding. The first inflection point near the K-edge, E_i, moves by 3 eV in going from a metallic solution (x = 0.05, 363 K) to the non-metal state (x = 0.03, 197 K). This indicates there is more electronic shielding in the metallic states than in the insulating state. The Rb-N distance as well as the solvation number derived from the X-ray absorption are given in Table I. The solvation number is n = 6, independent of the mole ratio $\mathcal{R}$ = (1-x)/x, and the Rb-N distance is ≈ 3.1 A, also independent of $\mathcal{R}$; although shorter mean values are observed at higher metal concentrations indicating there is short range order in solution.

The thermodynamic data must be reconciled with the above structural information. The chemical potential for $Na_x(NH_3)_{1-x}$ measured in the cell described in Table II [4] is written:

$$\mu = \mu\{Na_x(NH_3)_{1-x}\} - \mu^{\ominus}(Na) = \mu^{\bullet}_{free} + 2RT\ln(x/X) + \mu_{em} \qquad (1)$$

where x/X is the particle mole fraction which takes into account solvation with n = 6. The number of free solvent molecules is 1-7x so that X = 1-5x for a given x was used to evaluate μ_{ideal} in Figure 2. μ_{em} is the extra electromagnetic free energy per mole of metal above that for the ideal state. The latter is measured directly from the data in Figure 2 [4]. The physical significance of μ_{em} is obtained by comparing (1) with the equilibrium relations deduced by Schindewolf and Werner from their data in Table II, i.e., for a_s = 1:

$$\mu_{em} = 2RT\ln\gamma_{\pm} + RT\ln\{\alpha(1+\alpha)/2\} = \mu_e + \mu_m \qquad (2)$$

The first term, μ_e, represents the extra electrical work that can be derived from the extended Debye-Hückel treatment for $x < 10^{-3}$ and the second term, μ_m, represents the magnetic contribution which may be derived from the fraction of free spins, α, when Boltzmann statistics are obeyed by the system. μ_{em} is related to the excess free energy evaluated using the Gibbs-Duhem relation shown in Figure 3 [4]:

$$\Delta G^{Ex}_{Na} = (\mu - \mu_{ideal})_{Na} + \mathcal{R}(\mu - \mu_{ideal})_{NH_3} = \mathcal{R}\int_0^x \mu_{em}\, d\mathcal{R}^{-1} \qquad (3)$$

and the data in Figure 2 indicate that for $x < 10^{-4}$ to 10^{-3} $(d\mu_{em}/d\ln x) = -RT$ or $\mu_{em} = \Delta G^{Ex}_{Na} - RT$. However, the slope for the excess free energy versus x

$$d(\Delta G^{Ex}_{Na})/dx = x^{-2}\int_0^x \mathcal{R}^{-1}\, d\mu_{em} \qquad (4)$$

varies rapidly in this region where the extended Debye-Hückel ceases to be valid. Thus, the singularity in the slope of the excess free energy in Figure 3 must be due to rapid changes in the magnetic contribution μ_m near $x \approx 10^{-2}$ where the Knight shift, esr, and static susceptibility measurements indicate the solutions become diamagnetic [5]. A statistical analysis must take these phenomena into account.

3. Statistical Analysis of μ_{em}

The extra free energy over the ideal value in a degenerate electron system including correlation can be calculated using textbook methods. Screening of the Coulomb field is the best understood phenomenon. When Boltzmann statistics are obeyed, the Debye-Hückel solution of the Poisson equation obtains the screened potential:

$$V_{DH} = V_o \exp(- r/r_D,) \qquad (5)$$

where V_o is the Coulomb potential and the Debye distance:

$$r_D = (r_i^3/(6r_B))^{1/2} >> 0.5\ r_B = 8.4 \times 10^{-6}\text{meter}/(DT), \qquad (6)$$

varies as the inverse square root of the concentration ρ_o when $2r_i$ is the homogeneous distance between two ions in solution and the inequality for the Bjerrum distance, r_B, must be satisfied. In liquid ammonia (D = 22.4) this is obeyed only for $x < 10^{-3}$ near 240 K. Then the extended Debye-Hückel treatment:

$$\mu_e = \int_{\text{charge}} d(V_{DH} - V_o) = -2e^2/(4\pi\varepsilon(r_D + r_1)) \qquad (2')$$

gives the usual expression for the activity coefficients when ε is the permitivity and rl is a radius of nearest ion approach.

Fermi-Dirac statistics are obeyed near $x > x_c$. Then the Thomas-Fermi expression for the electron density in the Poisson equation gives:

$$V_{TF} = V_o \exp - r/r_o \qquad (5')$$

and

$$r_o = 0.5\ (\pi a_H/k_F)^{1/2} = r_D(E_F/(1.5k_BT))^{1/2} \qquad (6')$$

varies much more slowly than r_D versus x, as shown in Figure 4. a_H is the Bohr radius in the dielectric medium (a_H = 0.529 D in Angstrom) and $k_F=(3\pi^2\rho_o)^{1/3}$.

At sufficiently high temperatures ($E_F = \hbar^2k_F^2/2me \leq 1.5\ k_BT$), Boltzmann statistics are obeyed for all x. But for $x_c = 0.04$, E_F=0.39 eV [6] requires that $T > 3 \times 10^3$ K. Thus, near 240 K and $x > x_c$:

$$\mu_e = \int_{\text{charge}} d(V_{TF} - V_o) = -e^2/(4\pi\varepsilon r_o). \qquad (2'')$$

Figure 4 shows that $(r_1 + r_D) \rightarrow r_0$ so that the two approximations are similar. However, relations (2') and (2") apply only to phenomena where the difference in momentum between the initial and final states $\hbar q \rightarrow 0$ [7]. In metals, the largest contribution to the extra free energy arises when $q = 2k_F$, giving rise to what is known as Friedel oscillations in the free electron concentration which fall off as $r^{-3} \cos 2k_F r$ [7] from say an impurity center. These terms are the most important contributions to nuclear spin interactions in metals [8], and their presence in metal ammonia may be investigated as a function of the Knight shifts on k_F [5]. In the intermediate concentration region, $10^{-3} < x < x_c$, the impurity centers may be the sodium anions described in Table II where the observed diamagnetism [4] indicates that correlation terms dominate the interactions. The exact statistical treatment in this region is complicated, but a qualitative description is available.

According to Mott [1], a MI transition will occur in an array of one electron atoms as their separation increases because two electron correlation interactions in a rarefied gas lead to the formation of a non conducting phase [7]. This type of transition is called a Mott Transition. The non-metallic phase can be diamagnetic or antiferromagnetic, and Overhauser [9] has described these types of instabilities with $q = 2k_F$ by introducing a correlation potential, G, into the one electron Schrödinger Equation:

$$\{p^2/2me + G \cos (q\cdot r - \delta) - (E_k - V_0)\}|\, k > = 0 \qquad (7)$$

when E_k is the energy for the Bloch state, $|k>$. There are two types of instabilities which may occur in isotropic metals with a deformable uniformly charged background or "jellium". A spin density wave (SDW) can appear when there is a spatially inhomogeneous charge density for the two different spin states that are π out of phase. In this case, the charge oscillations cancel, leaving a spin polarized state. A charge density wave (CDW) is different from the SDW in two ways. The charges associated with the different spins are modulated in phase, giving rise to a diamagnetic state with a charge modulation, i.e.,

$$\rho = \rho_0[1 + p \cos (q\cdot r - \delta)] \qquad (8)$$

This is associated with a positive ion displacement of amplitude $a = p/q$ and δ is a phase shift. Thus, relation (8) represents the first term in the Friedel equation for the free electron concentration oscillations when $q = 2k_F$, i.e., [7],

$$\rho - \rho_0 = 1/(2\pi^2 r^3) \sum_{\ell = 0} (-1)\ (2\ell + 1) \cos (2k_F r - \delta_\ell) \sin \delta_\ell . \qquad (8')$$

and the Poisson equation is solved retaining only the first term in (8'):

$$\nabla^2 G \cos (2k_F r - \delta_0) = -e^2(\rho - \rho_0)/\varepsilon. \qquad (9)$$

Then, assuming a slow varying G and $r = yr_i$, obtains:

$$G = E_F(2/(\pi^2 y^3))\ (\sin \delta_0/((1 + 2 \tan(2k_F y r_i - \delta_0))(k_F a_H)). \qquad (10)$$

and $\mu_{em} = 1/V \int_{volume} G \cos(2k_F r - {}_0) dV.$ (10')

The phase shift, δ_0, is estimated using the Friedel sum rule for the difference in charge ΔeZ between the background ions Na^+ and the anion species, i.e., [7]

$$\Delta Z = -2 = -2 \Sigma \ (2\ell + 1) \delta_\ell / \pi, \quad (11)$$

gives $\delta_0 = \pi$ in the simple approximation which retains only the first term in (11). Then (10) is non zero only when the denominator also vanishes (for y --> 0.7) i.e.,

$$G \to -0.3 \ E_F/(k_F a_H) \text{ as } \delta_0 \to \pi. \quad (12)$$

This means that there is a concentration region where correlation effects leading to a diamagnetic state ($x = 10^{-2}$) are enhanced because the denominator in (10) vanishes.

4. Conclusions

The screening of the electrostatic field in metal ammonia solutions may be described by the simple Debye-Hückel treatment only when $x < 10^{-3}$ or by the simple Thomas-Fermi analysis when $x > x_c$. However, in the intermediate region where r_0, r_D, and $r_B/2$ are equal (Figure 4), the screening of the electrostatic field is poorly understood. In this region, charge density wave instabilities may be produced by the formation of anion species. The changes of entropy with pressure can lead to the separation of a phase such as $Na(NH_3)_6{}^+ \cdot Na^-$ reported in this meeting by Winter, Voronel and Schindewolf [9].

5. Acknowledgements

This work was done with partial support from grants by the following U. S. agencies: NSF-DMR 790011, NSF-DMR 77-27489 and DOE DE-AC03-76 SF 00098. Ms. S. Brown and Ms. H. Stein did the editing.

6. References

1. N. F. Mott, J. Phys. Chem. (1980), 84, 1199-1203.
2. J. V. Acrivos, K. Hathaway, A. Robertson, A. Thompson and M. P. Klein, ibid, (1980), 84, 1206-1211.
3. P. Chieux, P. Damay, J. Dupuy, and J. F. Jal, ibid (1980), 84, 1211-1215.
4. U. Schindewolf and M. Werner, ibid (1980), 84, 1123-1127.
5. J. V. Acrivos and N. F. Mott (1971) Philos. Mag., 24, 19-31.
6. J. C. Thompson, "Metal Ammonia Solutions", Lepoutre and Sienko, ed. (1964) Benjamin, 306-313.
7. D. J. Thouless, "Quantum Mechanics of Many Body Systems," 2nd ed. (1972).
8. M. A. Ruderman and C. Kittel, Phys. Rev. (1954), 96, 99.
9. R. Winter, A. Voronel, and U. Schindewolf, this meeting.

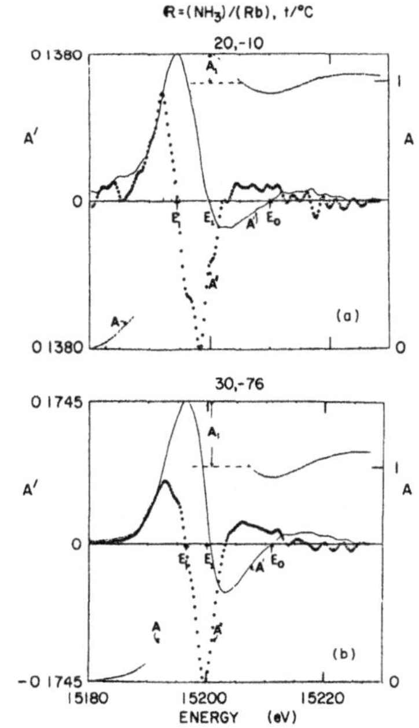

Figure 1.
X-Ray absorption near the Rb K-edge for MAS near x_c.

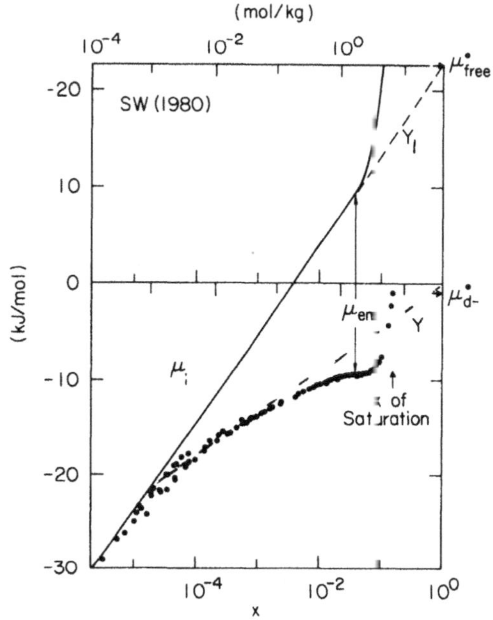

Figure 2.
Chemical potential measured by Schindewolf and Werne- [4] for the cell described in Table II at 238 K.. μ_1 is the ideal value calculated for solvated cations with n = 6. $y_1 = \mu^\circ_{free} + 2RT \ln x$ and $y_2 = RT \ln x$ gives $\mu_{em} = -\mu^\circ_{free} - RT \ln x$ for $10^{-5} < x < 10^{-3}$.

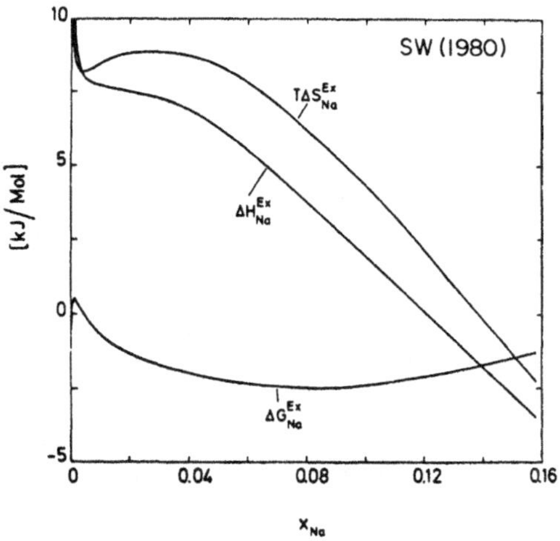

Figure 3

Excess mixing functions in MAS according to [4] vs x.

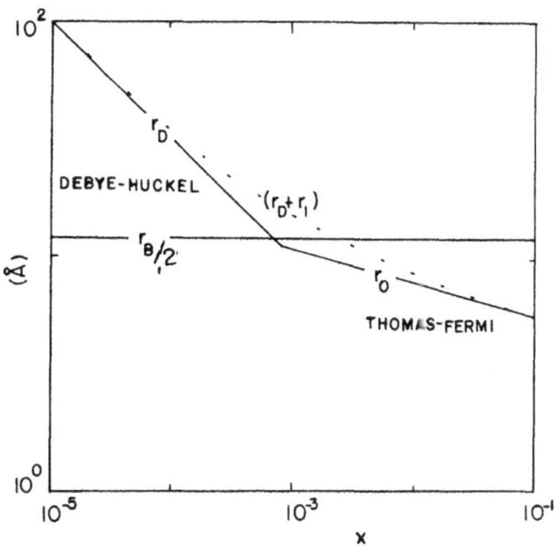

Figure 4.
Screening radii for MAS near 238 K. When $(r_D + r_1) = r_0 \approx r_B/2$, there are no satisfactory statistical models to describe the system.

Table I. X-Ray absorption data for Rb-NH_3 near the Rb K-edge near 200 K.

EXAFS Data		
R	Rb-N[(a)]/A	NH_3 Solvation N° ± 1
7	3.071 (6)	5.2
20	3.09 (5)	5.2
30	3.09 (7)	6.4
36	3.13 (5)	6.6
200	3.096 (10)	5.7

(a)The uncertainty in the last significant figure is given in parentheses.

Table II. Data for the Schindewolf and Werner Electrochemical Cell: Pt(c) | Na(c) | Na^+-β-Alumina | $Na_x\ (NH_3)_{1-x}$ | Pt(c) and their proposed cell reactions in reference [4].

$$Na(c) + nS \Longleftrightarrow eS^-_{n-} + NaS^+_{n+},$$

or

$$\mu = -EF = \mu^\bullet_{free} + RT\ ln\{(a_{f-})\cdot(a_{f+})/(a_s)^n\}, \qquad (SW1)$$

and

$$Na(c) + eS^-_{n-} + nS \Longleftrightarrow NaS^-_{n+n-}$$

or

$$\mu = \mu^\bullet_{d-} - 0.5\ \mu^\bullet_{free} + RT\ ln\{(a_{d-})/((a_{f-})(a_s)^n)\}. \qquad (SW2)$$

Components	Free Solvent S	Solvated Electron eS^-_{n-}	Solvated Cation NaS^+_{n+}	Diamagnetic Anion NaS^-_{n+n-}
mole fraction[(a)] x_i	(1-x(1+n))/X	xα/X	x(1+α)/(2X)	x(1-α)/(2X)
activity: $a_i = x_i\gamma_i$	a_S	a_{f-}	a_{f+}	a_{d-}
Standard State	Pure	α=x=1 infinitely dilute solution		α = 1/3
Standard [(b,c)] μ(238 K)	0 (by definition)	$\mu^\bullet_{free}$ = 21.9 kJ/mol		$\mu^\bullet_{d-} \approx$ - 1kJ/mol

(a) When $n = n_+ + n_-$, $X = 1 - x(n - 1)$. α is the fraction of unpaired spins.

(b) $\mu^\bullet_{d-}$ is deduced from the data of ref. [4] with $\mu\ (\alpha = 1/3) = \mu^\bullet_{d-} - 0.5\mu^\bullet_{free}$.

(c) $\mu_{ideal} = \mu^\bullet_{free} + RT\ ln\{x^2(1-x(n-1))^{n-2}/(1-x(n+1))^n\}$

The data in Table 1 indicates n = 6.

TETRAHEDRALLY COORDINATED IONIC LIQUIDS – ANOMALOUS TRANSPORT PROPERTIES, AND GEOPHYSICAL SIGNIFICANCE.

C. A. Angell, P. A. Cheeseman and S. Tamaddon
Department of Chemistry
West Lafayette, Indiana 47907
U.S.A.

The anomalous behavior of water with respect to both density-temperature relations and viscosity-pressure relations, is generally recognized, and its origin has been associated with the existence of an open network structure as the preferred low temperature (hence low energy) arrangement of the constituent particles. It is also generally known that the tetrahedrally coordinated inorganic compounds, BeF_2 and SiO_2, have crystalline characteristics in common with those of ice. It is therefore natural to consider the possibility that, in the liquid state, these substances and others related to them might manifest water-like anomalies. However, due to the high temperature regime in which these liquids must be studied, experiments to establish the possible existence of such relationships have not been performed.

In the case of silica, measurements on glassy structures at room temperature, obtained by quenching the liquid from different liquid range temperatures, have been able to establish the existence of a density maximum in the (supercooled) liquid state of this substance(1), which however is a weaker feature than in the case of water.(2) This observation encourages the search for viscosity-pressure anomalies, which in the case of silicates would be of great interest because of their relevance to geophysical problems, particularly planetary evolution. On a purely academic level, the observation of such liquid state anomalies in the bridge-bonded and relatively ionic materials would be interesting because of an implied geometrical origin, as opposed to hydrogen bonding per se, for the anomalous characteristics of water.

In this work we investigate the problem of high temperature ionic liquid behavior under pressure, by the relatively painless method of ion dynamics computer simulation.

Almost ten years ago, Rahman et al.(3) showed that a system of rigid ions would form a tetrahedral network structure when the ion size and charge parameters were suitably chosen. They simulated liquid BeF_2 and obtained excellent agreement with experimental structural features for this liquid (though the simulated system was somewhat more diffusive than laboratory BeF_2).

In subsequent work Woodcock et al.(4), somewhat more surprisingly, showed that the important structural and energetic features of liquid silica could also be reproduced by calculations utilizing centrally symmetric ionic pair potentials, in which the full formal charge of the ions were retained. This work demonstrated also, in preliminary form, that an increase of pressure on a system initially at ~ zero pressure, led to increases in the diffusion coefficient of the oxide ion, though the limited computer time available for these studies left the detailed behavior in some doubt.

In the meantime, laboratory studies of a more complex but experimentally more tractable material were reported by Kushiro[5], who demonstrated that when the pressure acting on liquid $NaAlSi_2O_6$ was increased to 40 kbars the viscosity at 1350° C decreased by about an order of magnitude. Since this substance has, an interesting additional feature, an alkali ion which resides within the pore structure of an otherwise tetrahedral network arrangement of Si, Al, and O cations, we decided to carry out an investigation by the ion dynamics method of the response of the system to increasing pressures which extend far beyond the range of feasible experimentation. Our interest was not only to demonstrate that the observed behavior was characteristic of a purely ionic system, but to determine the pressure at which a minimum in the viscosity might be encountered. In the simulation, a minimum viscosity would be indicated by a maximum in the diffusivities of the ionic species making up the network. Actual viscosity calculations are not practical at this time because, being related to n- particle correlation functions, they require much longer computing times.

The calculations are carried out on a system of 190 ions interacting pair-wise through simple potentials of the form

$$U_{ij}(r) = e^2 z_i z_j / r_{ij} + b_{ij} \exp [(\sigma_i + \sigma_j - r_{ij})/\rho]$$

for which the repulsion term parameters are listed in Table 1.

Table 1

Alkali Silicate and Alumino Silicate

Species	σ (Å)	b_{ij} (10^{-13}erg)			
		Si	O	Na	Al
Si	1.33	3.42			
O	1.42	2.117	0.814		
Na	1.252	2.898	1.595	2.375	
Al	1.358	3.372	2.070	2.850	3.325

$\rho = 0.29$

Production runs were extended up to a maximum of 5 picoseconds of which the last 4 were utilized for determination of the diffusivities defined by the limiting slopes of the mean-squared displacement time plots. After trial runs demonstrated that the slope of such plots changed little after the first 2 picoseconds, the production runs were cut back to 3 picoseconds in the interest of obtaining more points at different pressures in the available computer time. While the reliability of the calculated diffusivities is not considered high, we believe it is sufficient to ensure that the qualitative features of the results shown below are reliable.

To ensure that sufficient displacements of the ions occurred in this time period, the temperature of 6000 K was chosen for most of the investigation. At this temperature the lowest diffusivities are of order $5 \times 10^{-5} cm^2 sec^{-1}$. Unfortunately, direct comparison with the $NaAlSi_2O_6$ experiments, which were carried out at 1623 K, is out of the question, because the equivalent diffusivity is some 4 orders of magnitude beyond the present simulation capability. If the network structure has been destroyed by large concentrations of added oxide ion, however, then the experimental and simulation ranges approach each other and comparisons at 1500 K can be made.[9]

Results

The characteristics of the motion of network and non-network ions are depicted in Fig. 1 which shows the trajectories of all the ions of type Si^{4+}, and Na^+ during 400 time steps after projection onto one of the box faces. It can be seen that the displacements of silicon ions occur by a sequence of small and randomly directed motions which, however, feature frequent reversals in trajectories implying an effective, and rather high, "rattling" frequency. The same characteristics are found for the Al^{3+} and O^{2-} dynamics. The motion of the sodium ion is rather different, involving long uninterrupted flights of as much as half a box length. These presumably are directed along the open channels characteristic of the network structure. Print-outs of the frequency of trajectory reversal, however, show that for the sodium ion also, an effective rattling frequency exists, the frequency being about 260cm^{-1}. This is higher than the values reported earlier for $Na_2O \cdot 3SiO_2$ at 1500 K, 190cm^{-1}[6], and is close to that expected from the known far infra-red absorption spectrum of glassy silicates containing sodium ions, 230cm^{-1}. For 22 $Na_2O \cdot 53SiO_2$ Soules[7] has calculated a reversal frequency of 260cm^{-1}.

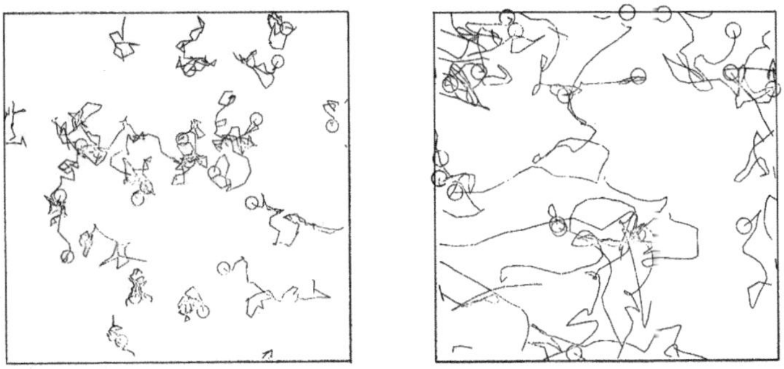

Figure 1. - Trajectories of all Si^{4+} ions (left) and Na^+ IONS (right) in the primary box during 400 time steps (0.8 ps), projected onto a box side.

The mean-squared displacements for each ionic species are displayed in Fig. 2 which shows that the sodium diffusivity is very much greater than that of any of the network ions at low pressure, but more comparable at high pressure.

In Fig. 3 we show how the diffusivities of these species, obtained from the slopes of the Fig. 2 plots, vary as the pressure is increased systematically up to values in the vicinity of 600 kbar.

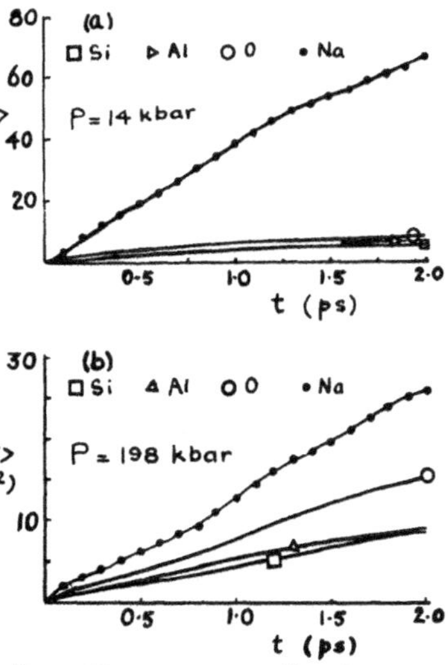

Figure 2. - Mean square displacement vs time for each species in system (a) low pressure and (b) intermediate pressure. (Individually Si, Al, and O points omitted for clarity).

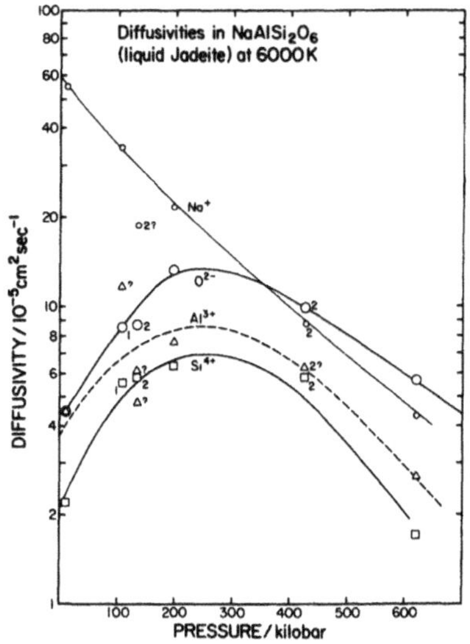

Figure 3. - Diffusivities of different species in liquid $NaAlSi_2O_6$ as a function of pressure, showing mobility maxima for network species at about 250 kbar.

DISCUSSION

The outstanding feature of Fig. 3 is, of course, the maximum in the diffusivity for each of the network-forming ions, which is encountered in the pressure range 200-300 kbar. Although the uncertainty on the individual diffusivities is rather high (note certain discrepant values for $D_{Al^{3+}}$) the overall behavior is rather convincing. By contrast the sodium diffusivity decreases rapidly and progressively at all pressures, undoubtedly because of the destruction with increasing pressure of the channels through which it could move so easily at low pressures.

In view of the fact that, in the case of water, increase of temperature above 25° C (1.1 T_m) is sufficient to eliminate the existence of the minimum viscosity phenomenon, it is at first sight surprising that the same phenomenon should be observable in the present systems at temperatures as high as 4 T_m. The persistence of the phenomenon in the case of silicates must reflect the strength of the oxide bridge-bond, and the relatively small configurational degeneracy of the resultant network. The contrasting "fragility" of the H_2O network structure is implied by the large excess heat capacity of the liquid compared with the crystalline form (C_p(water)/C_p(ice) = 1.64). This contrasts with the case of silica, and other oxide bridge-bonded networks which are characterized by unusually small excess heat capacities (C_p(liquid)/C_p(crystal) ≈ 1.05).

Although at this time studies have not been performed on liquid BeF_2 as a function of pressure, it seems highly probable that anomalous viscosity-pressure relations will be found in this case also. Unfortunately, it is not possible to make charge compensated replacements of the beryllium cation in fluoroberryllates analogous to the (Al^{3+} + Na^+) substitution for silicon in the SiO_2 structure.

In his simulations of the BeF_2 structure, Brawer[8] has noted that the occurrence of occasional 5 coordinated Be^{2+} ions in the structure is associated with an unusual level of F^- mobility in the vicinity. The implication that 5-coordinated network centers might be important in the occurrence of anomalous pressure dependences of mobility may be tested in the present instance by examining the coordination numbers, defined by the integral under the first peak in the Si^{4+}-O^{2-} pair distribution function. Indeed, as we have shown elsewhere[9], the maximum coincides with a maximum prevalence of 5 coordinated Si^{4+}, though it is not to be concluded that a symmetrical coordination group exists at this pressure.

GEOPHYSICAL SIGNIFICANCE

The occurrence of water-like anomalies in many liquid silicates of common cosmochemical compositions is of obvious geophysical and cosmophysical importance. At temperatures lower than those of this calculation, the proportional increase in fluidity with increasing pressure would become much larger than that observed in this work implying that molten planetary bodies may be much more fluid in the interior than at the exterior; further there would be an interior zone or maximum fluidity. Hydrodynamic modelling of planetary evolution will need to take such effects into account.

REFERENCES

1. R. Bruckner, Glastech. Ber., 37, 413-425 (1964).
2. C. A. Angell and H. Kanno, Science, 193, 1121 (1976).
3. A. Rahman, R.H. Fowler, and A.H. Narten, J. Chem. Phys. 57,3010 (1972).
4. L. V. Woodcock, C. A. Angell, and P. A. Cheeseman, J. Chem. Phys., 65, 1565 (1976).
5. I. Kushiro, J. Geophys. Res., 81, 6347-6350 (1976).
6. C. A. Angell, L. Boehm, P. A. Cheeseman and S. Tamaddon, Solid State Ionics, 5, 597 (1981).
7. T. F. Soules, J. Chem. Phys., 76, 969-975 (1981).
8. S. A. Brawer, J. Chem. Phys., 75, 3510-3521 (1981).
9. C. A. Angell, P. A. Cheeseman and S. Tamaddon, Science, Oct. 1982 (in press); Bull. Mineral., 1983 (in press).

INTERDIFFUSION IN A QUASI-IONIC LIQUID ALLOY

M. Soltwisch, D. Quitmann,
Institut für Atom- und Festkörperphysik
Freie Universität Berlin, D1000 Berlin 33

H. Ruppersberg,
Institut für Werkstoffwissenschaften
Universität des Saarlandes, D6600 Saarbrücken 15, and

J.B. Suck
Institut für angewandte Kernphysik II
Kernforschungszentrum D7500 Karlsruhe.

In a binary liquid, the average concentrations of components A and B (c_A, c_B with $c_A+c_B=1$) as well as temperature T and pressure p are usually fixed. Strong spatial fluctuations of the local concentrations around the mean occur in ionic liquids because the interaction potentials V_{AA}, V_{BB}, $_{AB}$ (spatially averaged) are very different: $(V_{AA}+V_{BB}-2V_{AB})/k_BT>>1$. The double Fourier transform of the space-time correlation between concentration fluctuations is

$$S_{cc}(q,\omega) = c_A c_B [c_A S_{AA} + c_B S_{BB} - 2\sqrt{c_A c_B}\, S_{AB}] \tag{1}$$

where $S_{AA}=S_{AA}(q, \omega)$ etc. are the usual scattering laws, see /BhaT 70/. $S_{cc}(q, \omega)$ is connected to the interdiffusion constant D^+ by a Kubo relation /BhaT 74/

$$D^+ = \pi \lim_{\omega \to 0} \omega^2 \lim_{q \to 0} \frac{1}{q^2} \frac{S_{cc}(q,\omega)}{S_{cc}(q)} \tag{2}$$

Under favourable conditions, concentration fluctuations can be measured by scattering techniques, especially well if the casttering lengths b_i cancel on the average: $\bar{b}=c_Ab_A + c_Bb_B = 0$; see /ChiR 80/ for a recent discussion. Such is the case in $^7Li_{0.80}\ ^{nat}Pb_{0.20}$ /RupS 76/, and we have performed a quasielastic neutron scattering experiment on this alloy at the ILL Grenoble, using a time of flight spectrometer.

In liquid $Li_{0.80}Pb_{0.20}$, $S_{cc}(q)$ has a sharp peak at q=1.5 $Å^{-1}$ and a very low thermodynamic limit

$$S_{cc}(q=0) = N\,k_BT/(\partial^2G/\partial c^2) \approx 0.05 \cdot S_{cc}^{id}(q) \tag{3}$$

which prove the existence of strong heterocoordination /RupS 75/. Both, amplitude and range of the modulation of concentration fluctuations are comparable to the situation found in molten salts, and the heterocoordination is indeed to be ascribed to charge transfer and a considerable contribution of ionic bonding /RupS 76/, although conductivity is still dominated by electrons. Effective charges are estimated as $Li^{+0.2\ \ldots\ 0.3}$ and $Pb^{-0.8\ \ldots\ -1.2}$ /Eva 82/. The values of $S_{cc}(0)$ from a recent neutron scattering study of $^7Li_{.80}Pb_{.20}$ /RupR 82/ are reported in table 1.

The experiment is described in detail elsewhere /SolQ 82/, /SolQ 81/. Suffice it to state that momentum transfers $q = 4\pi\ \lambda^{-1}\ \sin\Theta/2$ have been $0.2 < q\cdot\text{Å} < 2.5$, and energy transfers $\hbar\omega$ up to -1 meV...+16 meV

Table 1: Average concentration fluctuations $S_{cc}(q\to 0)/c_A c_B$ for liquid $^7Li_{0.80}Pb_{0.20}$. For details about the experiment see /RupR 82/.

T/K	$S_{cc}(0)/0.16$
995	0.052
1025	0.054
1075	0.066
1125	0.071
1225	0.075

(depending on q) have been studied, temperatures were 1023 K, 1098 K and 1173 K, the melting temperature being 983 K. Data have been corrected in the usual way for background, multiple scattering, absorption etc.; the energy resolution was 0.28 meV.

At small q, where $S_{cc}(q) = \int S_{cc}(q,\omega)\,d\omega \ll 1$, one observes almost exclusively incoherent scattering from ^7Li which was analyzed to yield the Li self diffusion constant D_{Li}. Here we used Ficks law which allows a simple extrapolation of the incoherent part to all q's measured. The values for D_{Li} which are presented in table 2 are found to lie in between those for pure liquid Li at its melting point, and the extrapolation of that value to the present temperatures (about $6\cdot 10^{-5}\,cm^2/s$ and $30\cdot 10^{-5}\ cm^2/s$, respectively). As a result derived from $\bar{D}$, eq. (6)

below, we mention right here that D_{Pb} is roughly 2.0 ... $2.5 \cdot 10^{-5}$ cm^2/s, about the same as in pure liquid Pb near the melting point. Fits to the coherent data have been performed starting from the hydrodynamic limit for $S_{cc}(q, \omega)$ which implies exponential decay of the correlation between concentration fluctuations /CohS 69/, /BhaT 74/

$$S_{cc}(q,\omega) = S_{cc}(q)\,\frac{\Gamma^+/2\pi}{\omega^2+(\Gamma^+/2)^2} \tag{4a}$$

It contains the interdiffusion constant via

$$\Gamma^+ = 2\hbar q^2 D^+ \tag{4b}$$

However, a single Lorentzian shape describes at best the low energy-side (the long time behaviour) of the spectra properly. That the spectral slope has to change to a faster decay than $1/\omega^2$, as is the case for a Lorentzian, follows from the second moment sum rule /BhaT 74/ $\int \omega^2 S_{cc}(q,\omega)\,d\omega = q^2 k_B T/m^*$ (m^*=reduced mass). For a Lorentzian the second moment is infinite and one may introduce a cutoff of the theoretical spectra. The energies where this cutoff occurs ly far away from the energies accessible in the present experiment.

In fig. 1 we give the results from the fits due to eq. (4). The resolution was taken into account by a convolution procedure. Fig. 2 gives

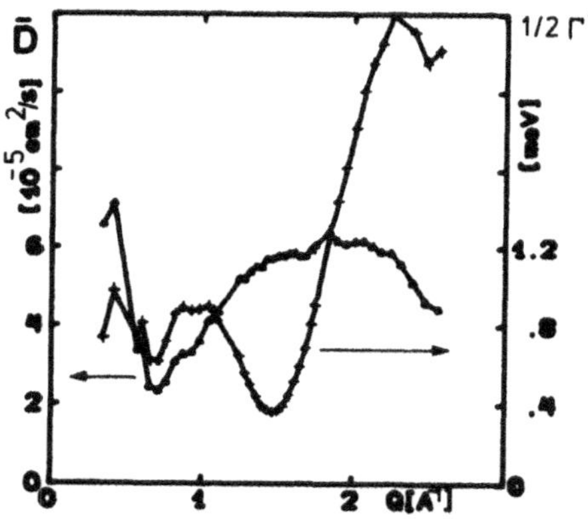

Fig. 1: Linewidths Γ^+ (crosses) and corresponding $\bar{D}=\Gamma^+ \cdot S_{cc}(Q)/c_A c_B \hbar q^2$ (triangles) from single Lorentzian fit (eqs. 4 and 5)

some examples of the fit-quality which is unsatisfactory below $q \approx 1\ \text{Å}^{-1}$. In fig. 1 the linewidth Γ^+ shows a pronounced q-dependence with a minimum at the structure maximum. It appears difficult to extrapolate to the hydrodynamic limit q=0, to obtain the proper transport coefficient.

From eq. (2) and (4) there is no explicit q-dependence for D^+ left, but the implicit q-dependence appears to be /SolQ 81/ of a form which is well known for q=0: There it simply expresses the fact that diffusional relaxation of concentrations is not driven by the gradient of the concentration, but of the chemical potential. See e.g. Tyrell's discussion of inter- and self-diffusion coefficients /Tyr 71/; Becker /Bec 79/ discusses the problem for D^+ in liquid Li_xPb_{1-x}, $x \leq 0.5$, based on work by Darken /Dar 48/. Extending this concept to finite q, one may expect

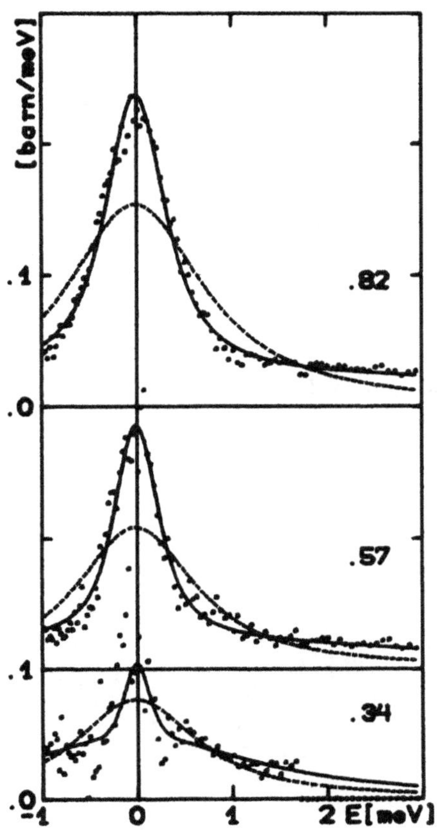

Fig. 2: Examples of the fit quality for three low q-values (0.34 Å^{-1} etc). Squares are the experimental 1023 K intensities after subtraction of the incoherent scattering from ^7Li. Solid lines: Continued fraction representation. Dashed lines: Hydrodynamic theory (eq. 4).

$$\bar{D} = D^{+}(q) \cdot \frac{S_{cc}(q)}{S_{cc}^{id}(q)} \approx const. \tag{5}$$

to hold. Here

$$\bar{D} = c_A D_B + c_B D_A \tag{6}$$

contains the individual self diffusion constants in the given binary liquid, and $S_{cc}^{id}(q) = c_A c_B$. The expectation is checked also in fig. 1. The strong dependence which $\Gamma^{+}(q)$ exhibits around $q=1.5$ Å^{-1} in fig. 1 is seen to cancel to a considerable degree, as expected from eq. (5). At the lowest q, where $\bar{D}$ rises steeply, the fit quality is bad.

As a description for $S_{cc}(q,\omega)$ which includes a priori the second moment rule, the authors have used a Mori-representation /SolQ 82/. D^{+} is then derived by comparing with the hydrodynamic form at $\omega=0$. The values thus obtained for $D^{+}(q)$ fulfill eq. (5) definitely better: $\bar{D}$ depends less on q and extrapolates for q=0 to a finite positive value, see fig. 3. No explanation is known to the present authors of the smooth rise of $\bar{D}$ for $q \lesssim 1.7$ Å^{-1}. Using $S_{cc}(0)$ vs. T from table 1 in eq. (5) we arrive at the last entry in table 2, which we believe to be the best estimate of D^{+}.

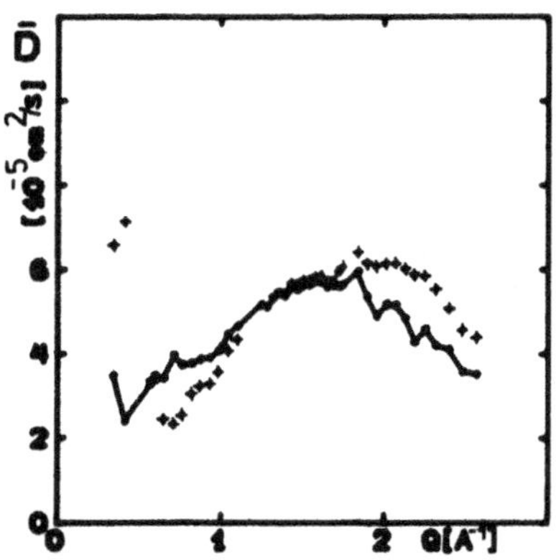

Fig. 3: Comparison of diffusion constants $\bar{D} = D^{+}(q) \cdot S_{cc}(q)/S_{cc}^{id}(q)$ for 1023 K. Circles correspond to fit results from a continued fraction representation, crosses are from the hydrodynamic form (eq. 4). Interdiffusion constants of table 2 were derived by extrapolating the continued fraction results to q=0.

Table 2: Diffusion constants in liquid $^7Li_{.80}Pb_{.20}$.

	1023 K	1098 K	1173 K	
D_{Li}	18 ± 2	21 ± 2	27 ± 3	10^{-5}
$\bar{D}(q\to 0)$	2.0 ± 0.5	2.7 ± 0.5	3.6 ± 0.5	$\frac{cm^2}{s}$
D^+	37	38	49	

This is, to our knowledge, the first study of the interdiffusion constant at finite q, i.e. on a microscopic scale, for a binary heterocoordinated liquid. It is felt that the general results like eq. (5) together with eq. (2) may well be valid in many heterocoordinated binary liquids. Application presupposes, however, knowledge or a good estimate of the individual diffusion constants in the same system. It should be realized also that for a truly ionic $A^+ B^-$ liquid, where $S_{cc}(0)=0$ exactly, eq. (5) cannot be valid at q=0; there, macroscopic concentration fluctuations would excite optical modes which in the present case play only a secondary role, see /SolQ 82/.

This work has been supported by Deutsche Forschungsgemeinschaft through Sfb 161. One of the authors (D.Q.) is obliged to Institut Laue-Langevin, Grenoble, for its hospitality.

REFERENCES

/Bec 79/ W. Becker, Dissertation Saarbrücken (79), unpublished

/BhaT 70/ A.B. Bhatia and D.E. Thornton, Phys. Rev. B2 (70) 3004

/BhaT 74/ A.B. Bhatia, D.E. Thornton and N.H. March, Phys. Chem. Liq. 4 (74) 97

/ChiR 80/ P. Chieux and H. Ruppersberg, J. de Physique 41 (80) C8-145

/CohS 69/ C. Cohen, J.W.H. Sutherland and J.M. Deutch, Phys. Chem. Liq. 2 (71) 213

/Dar 48/ L.S. Darken, Trans. AIME 175 (48) 184

/Eva 82/ R. Evans, private communication

/RupR 82/ H. Ruppersberg and H. Reiter, J. Physics F, in print

/RupS 76/ H. Ruppersberg and W. Speicher, Z. Natforsch. 31a (76) 47 and earlier references quoted there

/SolQ 81/ M. Soltwisch, D. Quitmann, H. Ruppersberg, and J.B. Suck, J. de Physique 41 (80) C8-167 and Phys. Lett. 86A (81) 241

/SolQ 82/ M. Soltwisch, D. Quitmann, H. Ruppersberg and J.B. Suck, to appear

/Tyr 71/ H.J.V. Tyrell in Diffusion processes, eds. J.N. Sharwood et al., Gordon and Breach 1971

MOTT TRANSITION IN LIQUID GOLD-ALKALI ALLOYS

F. Brouers, C. Holzhey, Freie Universität Berlin, Fachbereich Physik (WE5), 1000 Berlin 33, West Germany

J.R. Franz
Department of Physics, Indiana University, Bloomington, Indiana 47405, USA

W. Schirmacher
Technische Universität München, Physik Dept. E 13, D 8046 Garching, West Germany

Abstract

By incorporating calculated liquid pair correlation functions into a tight binding model for the electronic structure of binary liquid alloys, we have developed a self consistent theory of the electronic and atomic structure of liquid alloys. Model calculations for the gold-alkali systems show the existence of a gap in the density of states if the ratio of the electronegativity -difference to the mean overlap integral exceeds a certain value.

For alloy exhibiting a gap, a metal non-metal transition occurs for concentrations in the neighbourhood of stochiometry. This metal non-metal transition occurs for values of the atomic parameters lying between those of AuK and AuRb, in agreement with experimental results.

In the present paper we discuss how this Mott transition is related to electron correlations. The screening plays an important role and is governed by the electron density.

1. Introduction

In the last few years a number of liquid alloys have been discovered which, although formed from two metallic components exhibit non-metallic behaviour in a concentration range near stochiometry (see Enderly 1978, Hensel 1979 or Warren 1979 for a review). It seems to be clear now that the bonding type of some of these anomalous liquid alloys is ionic (or, at least, partly ionic).

Neutron diffraction experiment (Ruppersberg and Reiter 1981) show a preference for unlike nearest neighbours (chemical short range order). The high degree of ionicity is consistent with the large electronegativity difference between the two components. The liquid gold-caesium alloy has a metal non-metal transition at equi-atomic concentration. It has a difference of electronegativity of 1.7 on the Pauling scale. That liquid AuCs behaves like a molten salt has been demonstrated by Krüger et al. (1977). The liquid has in fact an optical gap of 1.4 eV (Münster and Freyland 1979) and its structure can be described by means of a model of charged hard spheres (Evans and Telo da Gama 1980). If one considers the family of liquid alkali metal-Au alloys one observes (Nicoloso et al. 1978) at equiconcentration a dramatic rise of the conductivity with decreasing alkali atomic number (from ionic conductivities of less than 5 Ω^{-1} cm^{-1} to values in the metallic range 10^3 - 10^4 $\Omega^{-1}cm^{-1}$). The conductivities of liquid RbAu and CsAu are in the ionic range.

An important question is why some systems that have large differences in electronegativity between components exhibit strong non-metallic behaviour and others do not. This question can only be answered by treating both the electronic and the atomic structure self-consistently because of the strong interdependence of chemical ordering and electron charge transfer.
A first attempt to calculate electronic structure in a manner consistent with the underlying chemical order of the atomic structure was formulated by the present authors (Franz, Brouers and Holzhey 1980, 1982, Holzhey, Brouers and Franz 1981). By minimizing the free energy with respect to the chemical short range order parameter, we were able to calculate the density of states, the charge transfer and the degree of local order in CsAu, gold-based alkali alloys and liquid alloys with a polyvalent component like Li_4Pb and Cs_3Sb, (Franz et al., 1982). In a more recent work (Holzhey, Brouers, Franz and Schirmacher 1982) we have incorporated the liquid structure in a more realistic way by introducing partial pair correlation functions. These are calculated

using a charged hard sphere model in the mean spherical approximation. By identifying the charge or the hard spheres with the amount of electronic charge transfer defined in our previous model, we are able to combine the two techniques.

This yields a powerful tool that permits, for the first time, calculation of both structural and electronic properties of a liquid binary alloy in a self-consistent way.

By allowing us to incorporate the liquid structure into the calculation, the new method also permits an improved treatment of the role played by screening in the inter-atomic part of the Coulomb interaction. This is the role that we want to emphasize in this paper.

2. Model for the Electronic States

When one is interested in the local properties of a binary alloy A_xB_{1-x}, one can start from a tight-binding Hamiltonian in the Wannier basis

$$H = H_o + T = \sum_i E_i \,|i\rangle\langle i| + \sum_{i\neq j} t_{ij}\,|i\rangle\langle j| \qquad (1)$$

The t_{ij}, non-zero only when i and j are nearest neighbours, are the hopping integrals. In the alloy A_xB_{1-x} the random atomic levels assume one of the two possible values, E_A and E_B depending on whether an atom of type A or B occupies the site i. The hopping integral can take the three possible values t_{AA}, t_{AB} and t_{BB} according to the occupation of sites i and j. The density of states N(E) is given in terms of the configurationally averaged diagonal Green function $\langle G_{ii}\rangle$

$$N(E) = -\frac{1}{\pi}\,\mathrm{Im}\;\langle G_{ii}(E+io)\rangle = x_AN_A(E) + x_BN_B(E)$$

where $N_i(E)$ is the partial density of states and x_i the concentration of the ith component. For solid alloys besides E_i and t_{ij} the other parameters characterize the atomic structure of the alloy: the composition x, the coordination number Z and the chemical short range order parameter σ. The partial coordination numbers Z_{ij} are given by $Z_{ii} = Z(x_i+x_j\sigma)$ and $Z_{ij} = Zx_j(1-\sigma)$ $(i = A,B;\ i \neq j)$. When short range order is present CPA methods are inappropriate and as a first approximation one uses the Cayley tree approximation (Falicov and Yndurain 1975, Brouers and Franz 1982). This leads to the following set of equations (Brouers, Lederer and Heritier 1973):

$$\langle G_{ii}\rangle = x_AG_A + x_BG_B \qquad (3)$$

$$G_i = (E-E_i-\Delta_i)^{-1} \qquad (4)$$

$$\Delta_i = Z_{ii}\,t_{ii}^2\,(E-E_i-\tilde{\Delta}_i)^{-1} + Z_{ij}\,t_{ij}^2\,(E-E_j-\tilde{\Delta}_j)$$

with $\tilde{\Delta}_i = \Delta_i\,(Z-1)/Z$

The effect of charge transfer is included by adding a local electron-electron contribution to the Hamiltonian depending on the redistribution of the electrons (charge transfer) in a self-consistent way:

$$E_i = E_i^o + \Delta E_i^{intra}(x,\sigma,U) + \Delta E_i^{inter}(x,\sigma,V) \qquad (6)$$

where U is the Hubbard intra-atomic and V the interatomic interaction term.

The short-range order parameters Z and σ that are normally used in the description of solid alloys are less well defined in the liquid state. The quantities that describe the structure of a binary liquid are rather the partial pair distribution function $g_{ij}(R)$.

For the calculation of the self energies Δ_i we let the coordination number Z go to infinity keeping the density ρ finite. This makes the distinction between Δ_i and $\tilde{\Delta}_i$ unnecessary. The electrons are now no longer restricted to nearest neighbour hopping and the summations over nearest neighbours are replaced by integrals over all spaces, weighted by the pair correlation functions $g_{ij}(R)$. Eq. (5) then takes the form (for i = A)

$$\Delta_A = \frac{\rho x_A \int t_{AA}^2(R)\, g_{AA}(R)\, d^3R}{E-E_A-\Delta_A} + \frac{\rho x_B \int t_{AB}^2(R)\, g_{AB}(R)\, d^3R}{E-E_B-\Delta_B} \tag{7}$$

In this new formulation the hopping integrals depend on the coordinate R. We assume that the wave functions are hydrogenoid s-state wave functions.

Once we have calculated the partial densities of states $N_i(E)$, we calculate the charge transfer

$$\delta n_i = n_i - n_i^o \tag{8}$$

where $n_i = \int_{-\infty}^{E_F} N_i(E)\,dE$ and n_i^o = number of valence electrons per atom in the pure metal i. The Fermi energy E_F is determined by the charge conservation equation. As before we replace the sums over nearest neighbour site by integrals over all space, using the partial pair correlation functions, and obtain for the inter-atomic energy shifts

$$\Delta E_i^{inter} = \delta n_i\, x_i\, \rho \int (g_{ii}(R) - g_{ij}(R)\; V(R)\; d^3R \tag{9}$$

To be consistent with the Hubbard type model which separates the Coulomb interaction into intra- and interatomic contributions we use the following interpolation formula for V(R) (Giner, Brouers, Gautier and Van der Rest 1976)

$$V(R) = \frac{e^2}{R+U^{-1}e^2} \tag{10}$$

3. Model for the Partial Structure Factors

To describe the structure of a liquid with partial ionic character we need a model which takes into account the charges of the ions. We have used the mean spherical approximation (MSA) (Waisman and Lebowitz 1972). In this scheme the interaction potential between an ion of type i with a charge $Q_i e$ and one of type j with charge $Q_j e$, a distant R apart, is given by

$$v_{ij}(R) = \infty \qquad R < d_{ij} \tag{11}$$

$$= \frac{Q_i Q_j e^2}{\varepsilon_o R} \qquad R > d_{ij} \tag{12}$$

ε_o is the dielectric constant (which we take equal to one) and d_{ij} the distance of closest approach between ions i and j.
The approximation invòlved in the MSA are best expressed in terms of the Ornstein-Zernike direct correlation function $c_{ij}(R)$ defined by

$$h_{ij}(R) = g_{ij}(R) - 1 \tag{13}$$

$$h_{ij}(R) = c_{ij}(R) + \sum_{k=i,j} n_k \int dR' c_{ik}(|R-R'|) h_{kj}(R') \tag{14}$$

The correlation functions $h_{ij}(R)$ being split into a pair term and a term due to the remaining atoms.

The asymptotic limit for large R is given by

$$\lim_{R\to\infty} c_{ij}(R) = -\frac{v_{ij}(R)}{k_B T} \tag{15}$$

The assumption of the MSA is that this relation holds at any range larger than the distance of closest approach

$$c_{ij}(R) = -\frac{Q_i Q_j e^2}{\varepsilon_o R\, k_B T} \tag{16}$$

for $R > d_{ij}$

while $\quad g_{ij}(R) = 0 \quad \text{for } R < d_{ij}$ (17)

The MSA simplifies considerably if one assumes that the hard sphere diameters are equal $d_i = d_j$. The neutrality condition imposes that

$$\sum_i x_i Q_i = 0 \tag{18}$$

Under these conditions the direct correlation functions $c_{ij}(R)$ can be obtained throughout all space. One can then calculate the partial structure factors from the expressions given by Ashcroft and Langreth (1967) for a binary alloy.

4. Introduction of the Screening into the Model

We also need to consider the fact that, if the liquid is not completely ionized, conduction electrons are present which screen the charged ions. The potential (11) used in the atomic structure calculations should now be a screened rather than a bare Coulomb interaction. Solutions of the MSA with a screened potential have been published by Waisman (1973) for a single component fluid. No complete calculations have been published yet for multi-component fluids. Preliminary numerical results for AuCs obtained by Copestake and Evans (1981) show that the partial pair correlation functions do not change appreciably when screening is introduced. Therefore as a first step, we have neglected the effect of screening on the pair-correlation functions.

In contrast, however, the effective interatomic Coulomb interaction term does change appreciably when conduction electrons are present. The phenomenological Coulomb interactions included in the tight-binding model used here should, therefore, be corrected to take into account the screening by the conduction electrons. Because we expect the screening to be strongly related to the value of the density of states at the Fermi level in the alloy, we have chosen to hold U constant and to introduce an exponential screening factor $e^{-\lambda R}$ in (10). Here $1/\lambda$ is the Thomas-Fermi screening length which is, as usual, proportional to the square root of the density of states at the Fermi level. This is included self-consistently in the calculations.

5. Self Consistent Calculation Scheme

The parameter Q_A now plays a similar role to that of the parameter σ in the solid state. If $Q_A = Q_B = 0$ the functions $g_{ij}(R)$ reduce to the solution of the hard sphere Percus-Yevick equation and are all equal. This means that there is no chemical order in the system. If $Q_A = -Q_B = 1$ the charge transfer is total and the system is salt-like with strong heterocoordination. Since Q_i represents the amount of charge polarisation in the system we identify this quantity with the charge transfer calculated from the density of states using Eq. (8)

$$Q_i = \delta n_i$$

From this condition the atomic structure parameter Q_i is directly related to the electronic structure in a self-consistent way. At equi-concentration one can start the self-consistency with a completely ionized liquid or with a completely neutral liquid. This leads to the same result.
The density of states at the Fermi level and the charge transfer for the whole family of gold-alkalis at stochiometry have been calculated in Holzhey et al. (1982). One can find in that paper as well as in Holzhey (1982) a discussion of the value of the various parameters. Let us say only that we have taken a hard sphere diameter d = 3.15 Å and that we have normalized $t_{ij}(R)$ such that

$$t_{ij}(R) = t_{ij}\ e^{-\alpha(R-d)} \quad \text{with} \quad \alpha^{-1} = d/2.$$

The results are listed in Table 1.

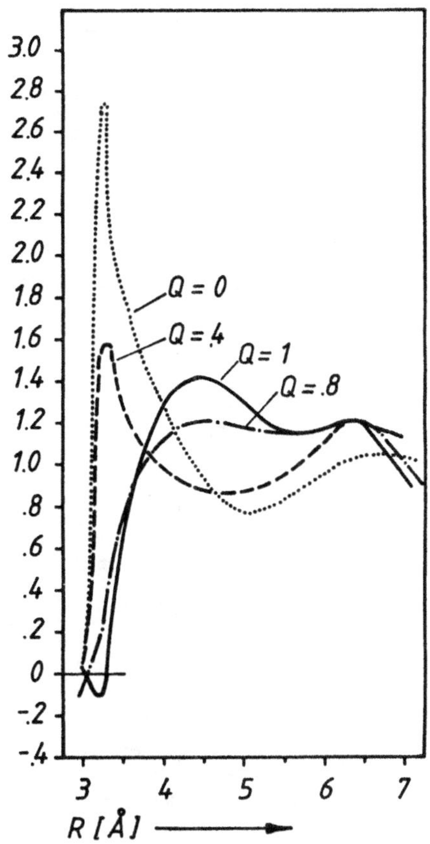

Fig.1 Partial pair correlation functions $g_{--}^{++}(R)$ calculated with the MSA-equations for different values of the charge Q on the ions.

It is of interest to consider the variation of $g^{++}_{--}(R)$ as a function of Q (Fig.1). There is a qualitative difference between the curves for Q = 1 and Q = 0.8 which corresponds to a molten salt situation and the curves for Q = 0 and Q = 0.4 which correspond to a metallic situation.

6. Discussion

In trying to establish a criterion for the opening of the gap (metal non-metal transition), we noted that only two parameters are specific to the particular alloys. These are $\Delta E = E_A - E_B$ and $\langle t \rangle$ the average hopping integral defined by $\langle t \rangle = \frac{1}{2}(t_{AA}+t_{BB})$. We have thus introduced the ratio $\Delta E/\langle t \rangle$ proportional to the electronegativity difference divided by an effective band-width to account for the difference between the alloys. This ratio increases monotonically from AuLi to AuCs. In Fig.2 we show the density of states $N(E_F)$ and the width of the gap for cases when $N(E_F) = 0$ as a function of $\Delta E/\langle t \rangle$. It can be seen that a very sharp transition occurs at a critical value of $\Delta E/\langle t \rangle = 5$, lying between the values for AuRb and AuK.
Because the electronegativity difference (and therefore ΔE) does not vary much from AuCs to AuLi, it cannot be the essential factor giving rise to the different electronic properties in this series of liquid alloys as this has sometimes been written (Nicoloso et al. 1978). It thus appears that the average overlap integral controls the transition.

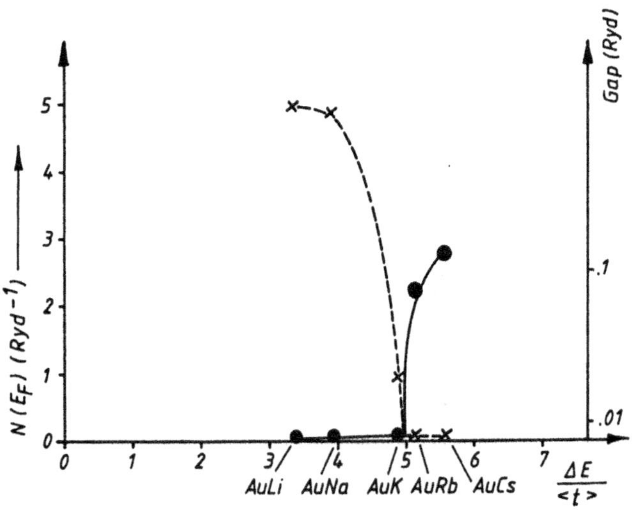

Fig.2 Density of states at the Fermi level $N(E_F)$, and width of the gap for cases when $N(E_F) = 0$ as a function of $\Delta E/\langle t \rangle$.

In our model we have defined the bandwidth of the alkali component as two times the Fermi energy. Therefore $\langle t \rangle$ is directly related to the electron density (see r_s the interelectronic distance of the alkali component in Table 1) which increases from AuCs to AuLi.

In LiAu the Coulomb interaction is more screened than in CsAu. This should not be surprising since due to the difference in electron density and consequently Fermi level and bandwidth, the Thomas Fermi wave vector is smaller in Cs than in Li and therefore in CsAu than in LiAu. The Coulomb potential being less screened in CsAu it is able to produce an ionic gap at equiconcentration. The Fermi level lies in a region of low density of states and this low density of states reinforces the difference of screening between Cs and Li alloys. This self-consistency serves as a feed back mechanism causing more and more charge transfer in the case $\Delta E/\langle t \rangle > 5$ or helping to close the gap in the case $\Delta E/\langle t \rangle < 5$. The liquid CsAu becomes an insulator if one discards the small ionic current while LiAu remains a good conductor for all concentrations. The other gold-based liquid alloys have a behaviour ranging between these two extreme cases following the difference in electron density.

The fact that the difference in electron density is the main factor determining the difference of behaviour of these liquids can be better understood if one views the phenomenon as a Mott transition due to electron correlation (Ziman 1964).

The distance between positive and negative ions in the liquid is $d_{+-} = 3.5$ Å according to the calculation of $g_{ij}(R)$ for $Q = 0.8$ (Fig. 1). If the screening length is smaller than this distance the system is metallic. This condition reads

$$\lambda > 1/d_{+-}$$

From Table 1 one can see that for AuK λ is a bit larger than $1/d_{+-}$ and that the transition occurs between AuRb and AuK for a critical electron density

$$n_c \simeq 0.036 \text{ Å}^{-3}$$

To observe the transition and its possible abruptness, it would be of interest to consider ternary alloys. However, it is to be expected that miscibility problems will obscure the phenomenon.

Outside stochiometry measurements of the electrical conductivity (Hoshino et al. 1975) and the magnetic susceptibility (Freyland and

Steinleitner 1976) indicate that for small concentration of excess cesium the electrons are localized. Recently NMR studies in Cs-CsAu solutions (Dupree et al. 1980) have shown that abrupt localization takes place for excess metal concentration below about 7 %. Our model can account for such an observation. Dupree et al. (1980) have interpreted their data by assuming the excess electrons to be trapped in states similar to F-centers in alkali halides. Littlewood (1981) has shown that this model can account for the localization of electrons observed close to the stochiometry composition. As the radius of the F-center is about the same as the radius of our charged spheres the Mott transition occurs for the same electron density. We therefore believe that in the liquid the real state should lie between the two extreme cases: the charge concentrated on the gold atom (which is the situation for exact stochiometry) and the charge forming a F-center for excess of Cs. On the other side of the transition where less information is available, other types of impurity centers should be considered.

Acknowledgements

This work is partly supported by the Deutsche Forschungsgemeinschaft and by a Nato grant.

We are grateful for helpful discussions with R. Evans and W. Schirmacher.

	AuCs	AuRb	AuK	AuNa	AuLi
t_{AA}	0.053	0.053	0.053	0.053	0.053
t_{BB}	0.017	0.020	0.023	0.035	0.045
t_{AB}	0.041	0.042	0.041	0.043	0.049
$\Delta\chi$	1.7	1.6	1.6	1.5	1.4
ΔE	0.198	0.186	0.186	0.174	0.163
$N(E_F)[Ryd^{-1}]$	0	0	0.9	4.85	4.86
gap[Ryd]	0.11	0.09	0	0	0
Q_A	0.8	0.77	0.66	0.44	0.38
$\lambda[\text{Å}^{-1}]$	0	0	0.53	1	1.2
r_s/a_o	5.62	5.20	4.86	3.93	3.25

Table 1. Model parameters for the Au(A)-alkali (B)-alloys. t_{ij} and ΔE are given in Rydbergs, $\Delta\chi$ is the difference of Pauling electronegativities. Results of the self-consistent calculations for Au-Alkali family at stochiometry. r_s/a_o is the interelectronic distance of the Alkali component.

References

Ashcroft N W and Langreth D C 1967 Phys. Rev. 156 685-692

Brouers F, Lederer P and Heritier M 1973 Amorphous Magnetism ed. H O Hooper and A M de Graaf (N.Y. Plenum) pp 151-8

Brouers F and Franz J R will be published

Copestake A and Evans R 1981 private communication

Dupree R, Kirby D J, Freyland W and Warren W W Jr 1980 Phys. Rev. Letters 45 130-33

Enderby J E 1978 The Metal Non-Metal Transition in Disordered Systems (Proc. 19th Scottish Univ. Summer School in Physics) ed. L R Friedman and D P Tunstall (Edinburgh: University of Edinburgh)

Evans R and Telo da Gama M M 1980 Phil. Mag. B 41 351-56

Falicov L M and Yndurain F 1975 Phys. Rev. B 12 5664-75

Franz J R, Brouers F and Holzhey C 1980 J. Phys. F: Met. Phys. 10 235-52

Franz J R, Brouers F and Holzhey C 1982 will be published

Freyland W and Steinleitner G 1976 Ber. Bundesges. Phys. Chem. 80 810-5

Giner J, Brouers F, Gautier F and van der Rest J 1976 J. Phys. F Met. Phys. 6 1281-89

Hensel F 1979 Adv. Phys. 28 555-94

Holzhey C, Brouers F and Franz J R 1981 J. Phys. F: Met. Phys. 11 1947-53

Holzhey C, Brouers F, Franz J R and Schirmacher W 1982 will be published in J. Phys. F

Holzhey C 1982 Doktorarbeit F.U. Berlin

Hoshino H, Schmutzler R W and Hensel F 1975 Phys. Lett. 51 A 7

Krüger K D, Fischer R and Schmutzler R W 1977 Inst. Phys. Conf. Ser. 30 480-7

Littlewood P B 1981 Phys. Rev. B 24 1710-14

Münster P and Freyland W 1979 Phil. Mag. B 39 93

Nicoloso N, Schmutzler R W and Hensel F 1978 Ber. Bunsenges. 82 621

Ruppersberg H and Reiter W 1981 will be published

Waisman E 1973 Molec. Phys. 25 45-48

Waisman E and Lebowitz J L 1972 J. Chem. Phys. 56 3086-93; 3093-4309

Warren W W jun. 1979 "Advances in Molten Salt Chemistry", Vol.4 Ed. J Braunstein et al. (Plenum Press)

Ziman J M 1964 Principles of the theory of solids, Cambridge University Press p.144

Schirmacher W 1980 J. Non-Cryst. Solids, 35-36 1301-6

SELF-MOTION IN CHARGED LIQUIDS

T. Munakata
Department of Applied Mathematics and Physics,
Kyoto University, Kyoto 606, Japan
and
J. Bosse
Institut für Theorie der Kondensierten Materie,
Freie Universität Berlin, D-1000 Berlin, Fed.Rep. of Germany

1. Introduction

Among various time-correlation functions the velocity autocorrelation function(VAF) is the simplest one which reveals properties of single-particle motion in condensed matter . Recently computer simulation studies have clarified detailed features of single-particle motion in charged liquids. In the classical one-component plasma(OCP) it was found[1] that the spectrum of the VAF has a sharp peak around the plasma frequency ω_p , suggesting strong coupling of single-particle motion to collective charge-density excitations(plasmon modes). The VAF of molten salts was also studied by molecular dynamics[2,3]. For the symmetric molten salt(SMS)[2], the VAF was found rather similar to that of liquid argon showing no pronounced oscillations. At present there is no microscopic theory to explain the difference between the VAF of the OCP and that of SMS and this report is devoted to the problem.

2. Formal Framework and Mode-Coupling Approximation

We consider a system consisting of S different ionic species $s=1,..,S$ with mass m_s ,charge e_s and number of ions N_s.($\Sigma N_s=N$, $\Sigma e_s N_s=0$, $n=N/V$) Let us assume the presence of a further tagged ion of mass m_0 and charge e_0 with position $\vec{r}_0(t)$ and momentum $\vec{p}_0(t)=\vec{v}_0(t)m_0$. The VAF is[4]

$$\psi(t)=\langle\vec{v}_0(t)\cdot\vec{v}_0\rangle/\langle[v_0]^2\rangle=(\hat{v}_0^{\,x}|\exp(-itL)|\hat{v}_0^{\,x}) \tag{1}$$

where the inner product $(A|B)$ is defined by $\langle\delta A^*\delta B\rangle/k_BT$ with $\delta A=A-\langle A\rangle$, $A(t)=\exp(itL)A$ and a normalized variable $\hat{A}=A/(A|A)^{1/2}$. The VAF obeys[5]

$$\psi(t)= -\int_0^t dt'K(t-t')\psi(t'), \quad \psi(t=0)=1, \tag{2}$$

$$\psi''(\omega)=(1/2)\int dt\, e^{it\omega}\psi(t)=K''(\omega)/[(\omega+K'(\omega))^2+(K''(\omega))^2] \tag{2'}$$

with the memory function

$$K(t)=(1/3)\Omega_E^{\,2}\sum_{\alpha}\quad(\hat{F}_0^{\,\alpha}|e^{-itQLQ}|\hat{F}_0^{\,\alpha}) \tag{3}$$

Here Q is the projection operator orthogonal to $\vec{v}_0$ and $\vec{F}_0$ is the force on the tagged ion $(\alpha = x,y,z)$:

$$F_0^{\,\alpha}=im_0Lv_0^{\,\alpha}=(1/iV)\Sigma_k\Sigma_s\; u_{0s}(k)k^{\alpha}e^{i\vec{k}\cdot\vec{r}_0}\; n_s(\vec{k}) \tag{4}$$

with $n_s(\vec{k})$ and $u_{0s}(k)$ denoting the Fourier transforms of the partial density $n_s(\vec{r})=\Sigma\delta(\vec{r}-\vec{r}_{j,s})$ and the interionic potential between the tagged ion and an ion of species s, respectively. Introducing a set of density variables $\rho_\sigma(\vec{k})$ $(\sigma=1,..,S)$ by $\rho_\sigma(\vec{k})=\Sigma_s n_s(\vec{k})A_{s\sigma}(k)(N/N_s)^{1/2}$, where $A_{s\sigma}$ represents the unitary matrix which diagonalizes the symmetric matrix of partial structure factors,

$$S_{ss'}(k)=\langle\delta n_s(\vec{k})^*\delta n_{s'}(\vec{k})\rangle/(N_sN_{s'})^{1/2} \;, \tag{5}$$

eq.(3) can be rewritten as

$$K(t)=(\Omega_E^{\,2}/3)\Sigma_\alpha\Sigma_{\sigma,\sigma'}\Sigma_{k,k'}(\hat{F}_0^{\,\alpha}|\hat{B}_\sigma(\vec{k}))(\hat{B}_\sigma(\vec{k})|e^{-itQLQ}|\hat{B}_{\sigma'}(\vec{k}'))(\hat{B}_{\sigma'}(\vec{k}')|\hat{F}_0^{\alpha}) \tag{6}$$

with $B_\sigma(\vec{k})\equiv e^{i\vec{k}\cdot\vec{r}_0}\;\rho_\sigma(\vec{k})$. The Einstein frequency is simply given by the two-body distribution functions[5].

A useful approximation suggested by Eq.(6) consists of the factorization of the normalized two-mode correlation function,

$$(\hat{B}_\sigma(\vec{k})|e^{-itQLQ}|\hat{B}_{\sigma'}(\vec{k}'))\simeq\delta_{\vec{k},\vec{k}'}\phi_0(k,t)\phi_\rho(\sigma,\sigma',k,t) \tag{7}$$

with the self-correlation function $\phi_0(k,t)=\langle e^{i\vec{k}\cdot(\vec{r}_0(t)-\vec{r}_0(0))}\rangle$ and the density correlation function $\phi_\rho(\sigma,\sigma',k,t)=(\hat{\rho}_\sigma(\vec{k})|e^{-itL}|\hat{\rho}_{\sigma'}(\vec{k}))$. Then the approximation (7) yields

$$K(t)=(\Omega_E^{\,2}/6\pi^2 n)\,\Sigma_{\sigma,\sigma'}\int dk\; k^2 W(\sigma,k)^* W(\sigma',k)\,\phi_0(k,t)\,\phi_\rho(\sigma,\sigma',k,t) \quad (8)$$

The vertex function $W(\sigma,k)$ is expressible in terms of the structure factors, Eq.(5) and their inverse, i.e. the direct correlation functions.

3. Application to the OCP and the SMS [6]

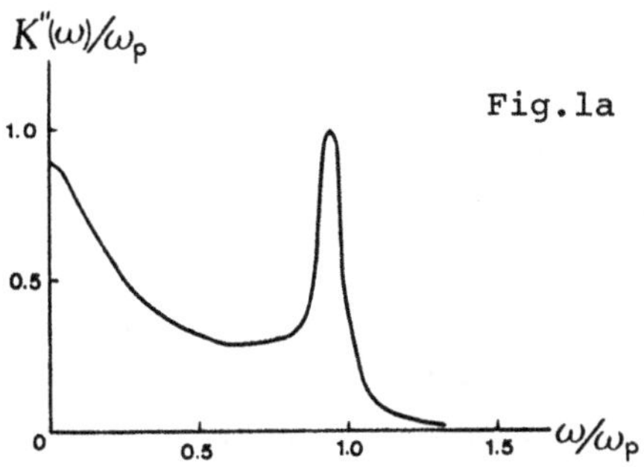

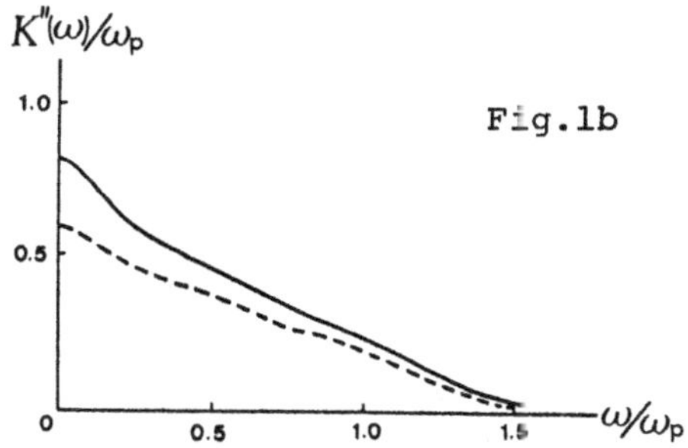

Fig.1a. Spectrum of the memory function for the OCP.

Fig.1b. Spectrum of the memory function for the SMS (full line). Dashed line: The contribution $K''_Q(\omega)$ from the term $\sigma=Q$ (Eq.(10))

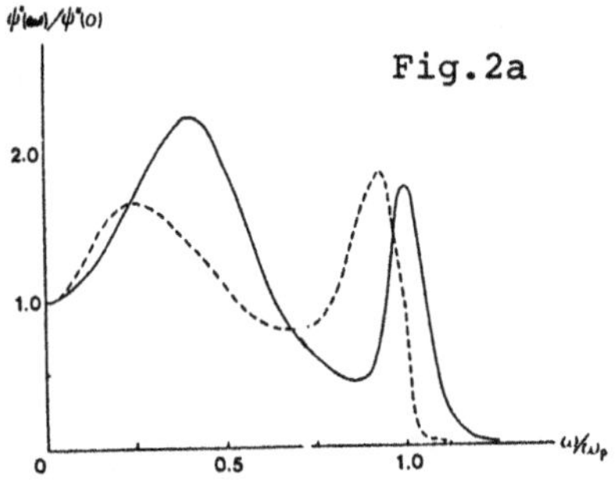

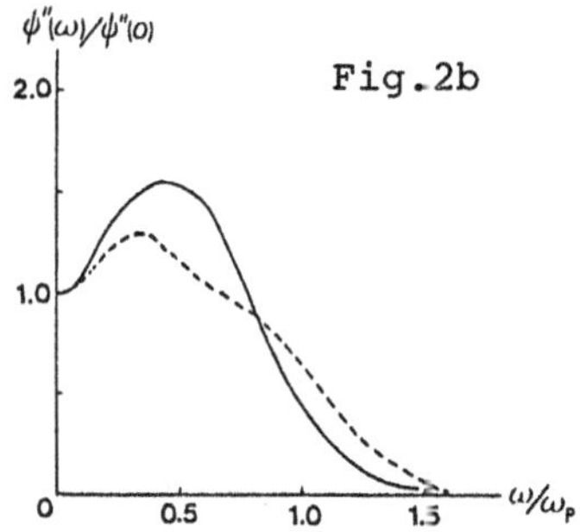

Fig.2 Normalized velocity autocorrelation spectra for (a) the OCP and (b) the SMS. Full line: present theory ($\Gamma=140$ for the OCP) Dashed line: computer simulation ($\Gamma=152.4$ for the OCP [Ref.1] and [Ref. 2] for the SMS).

For the OCP(S=1), it follows from Eq.(8) that

$$K''(\omega)=(\pi\Omega_{ocp}^{2}/I_{ocp})\int_0^\infty dk[k^2c(k)]^2\int d\varepsilon\ S_0(k,\varepsilon)S(k,\omega-\varepsilon) \tag{9}$$

with $I_{ocp}=\int_0^\infty dk[k^2c(k)]^2S(k)$ and $\Omega_{ocp}^{2}=\omega_p^{2}/3,(\omega_p^{2}=4\pi ne^2/m)$. The direct correlation function $c(k)$ is defined by $c(k)=1-1/S(k)$. The functions $S_0(k,\omega)$ and $S(k,\omega)$ denote the incoherent and the coherent scattering laws[4]. We note that the expression (9) is similar to the result $K_{GM}''(\omega)$ obtained by Gould and Mazenko,[see] 6 .One has $K(t)=(\omega_p^{2}/3)K_{GM}(t)/K_{GM}(t=0)$, i.e. our present approach conserves the correct value for $K(t\to 0)$. For the SMS Eq.(8) yields

$$K''(\omega)=(\pi\Omega_{sms}^{2}/I_{sms})\sum_{\sigma=M,Q}\int_0^\infty dk[k^2c_\sigma(k)]^2\int d\varepsilon S_0(k,\varepsilon)S_\sigma(k,\omega-\varepsilon) \tag{10}$$

where $I_{sms}=\sum_{\sigma=M,Q}\int_0^\infty dk[k^2c_\sigma(k)]^2S_\sigma(k)$ and $S_M(k,\omega)$ and $S_Q(k,\omega)$ denote the mass- and the charge-density correlation functions. Based on Eqs. (9),(10) and the Gaussian approximation for $S_0(k,\omega)$ [4], we calculated the spectrum $K''(\omega)$ with all necessary input data taken from computer simulation work[1,2]. (Fig.1). The spectrum of the VAF is obtained from Eq.(2') with $K'(\omega)$ connected with $K''(\omega)$ through the Kramers-Kronig relation. (Fig.2). These results show that the main features of the VAF's of the OCP and the SMS are reproduced well by the present theory. As to the diffusion constant our theory gives slight underestimation, i.e. $R=D_{theor.}/D_{exp.}=0.7$ for the OCP and $R=0.94$ for the SMS. While the memory spectrum of the OCP shows a sharp maximum close to the plasma frequency(Fig.1a) the spectrum of the SMS does not; it is a monotonously decreasing function of ω(Fig.1b). This qualitative difference in the memory spectra has the following reasons: (a) while the OCP has a sharply defined plasma mode for small k, this excitation is strongly damped in the SMS[2] (b) the dispersion of the plasma mode is much stronger in the SMS than in the OCP, and (c) the diffusion constant D is larger in the SMS than in the OCP, i.e. for small k $S_0(k,\omega)$ is narrower in the OCP. From these facts the difference in the memory spectra is easily understood. In Fig.1b we separately indicate the contribution $K''_Q(\omega)$ to $K''(\omega)$ which is due the coupling between single-particle motion and charge-density fluctuation.

Finally we comment on the mode-coupling(MC) approximation presented in Sec.2. We note that Eq.(8) can be derived via quite a different route, that is, through a linear response theory. The outline of the derivation is as follows[7]: We first calculate the number- and charge-density responses of other ions to some motion of the tagged ion, $\{r_0(t), \vec{v}_0(t)\}$ with the aid of a linear response method. Then we determine self-consistently the motion of the tagged ion which moves in the (mean) potential field $V_{m.f}(\vec{r},t)$ produced by the responses. In these steps of calculation we replace the bare interionic potential between the tagged ion and other ions by an effective one, which is specified so as to produce the correct static correlations under the linear approximation[8]. These procedures yield Eq.(8). Thus the vertex function $W(\sigma,k)$ turns out to correspond to the effective potential just mentioned and the $\phi_0(k,\omega)$ and $\phi_\rho(\sigma,\sigma',k,\omega)$ of Eq.(8) are interpreted , in the framework of the later approach, as expressing propagation of the tagged ion and the response of the surrounding ions, respectively.

Acknowledgment

This work is part of the research program of the Sonderforschungsbereich 161 supported by the Deutsche Forschungsgemeinschaft. One of us (T.M) acknowledges hospitality at the Freie Universität Berlin during his stay in Berlin.

References

1. J.P.Hansen, I.R.McDonald, E.L.Pollock, Phys.Rev. A11, 1025 (1975)
2. J.P.Hansen, I.R.McDonald, Phys.Rev. A11, 2111 (1975)
3. G.Cicotti, G.Jacucci, I.R.McDonald, Phys.Rev. A13, 426 (1976)
4. P.A.Egelstaff, An Introduction to the Liquid State (Academic Press, London and New York, 1967)
5. J.Bosse, W.Götze, A.Zippelius, Phys.Rev. A18, 1214 (1978)
6. T.Munakata, J.Bosse, Phys.Rev.A (1982) to be published.
7. T.Munakata, Prog.Theor.Phys. to be published.
8. J.K.Percus, The Equilibrium Theory of Classical Fluids, ed. H.L. Frisch and J.L.Lebowitz (Benjamin, New York 1964).

PROPERTIES OF Na/S AND LiAl/FeS HIGH-PERFORMANCE BATTERIES WITH MOLTEN SALT ELECTROLYTE

Reinhard Knödler

Brown, Boveri & Cie AG
Zentrales Forschungslabor
Heidelberg, F.R.G.

INTRODUCTION

The substitution of crude oil as a primary energy source has led to the development of new rechargeable batteries for electric vehicle traction and peak energy generation ("load levelling"). The two most promising candidates for these purposes are the sodium/sulfur ("Na/S") and the lithium aluminum/iron sulfide ("LiAl/FeS") systems. Both batteries function with molten salts at elevated temperatures. They offer the advantage of a considerable weight reduction compared to the conventional lead acid battery. In addition, cost benefits can be expected.

Work on the Na/S system is being conducted mainly in Germany (Brown Boveri), the United States (Ford, General Electric) and England (Chloride). The LiAl/FeS system is being developed in the United States (Argonne National Laboratory, Gould, Eagle-Picher) and Germany (Varta). Both batteries are similar in their performance data. However, in order to decide which battery will be most suitable for a specific application, a detailed analysis is necessary. This is presented in the following. The Na/S data reflect the Brown Boveri system, the LiAl/FeS data the Argonne, Gould and Eagle-Picher systems. The present study was conducted during a six month stay at Argonne National Laboratory. Thus, the assessment of the two battery types is based on direct experimental evidence and not merely on reports and publications which usually tend to be somewhat optimistic.

Na/S SYSTEM

The negative active material is sodium, the positive is sulfur. The sodium ion conducting ceramic β-Al_2O_3 serves as separator between these two liquid phases. Sodium is contained in a separate container inside the β-Al_2O_3 tube and sulfur is immobilized in graphite felt outside the tube. The graphite felt also serves as electronic conductor, where the electrochemically active species can react. The cells with a 50 Ah capacity are completely sealed and can be operated in open air. The operating temperature is 300 to 400°C. During discharge sodium polysulfides are formed according to the following reactions :

$$2\ Na \rightarrow 2\ Na^+ + 2e^-$$

$$2\ Na^+ + 5S + 2e^- \rightarrow Na_2S_5$$

This reaction is valid to a discharge depth of about 60% and is called the two-phase region, because Na_2S_5 and S exist in the melt. The open circuit voltage is here 2.08 V. If the cell is discharged further, Na_2S_4 and Na_2S_3 are formed successively :

$$4\ Na \rightarrow 4\ Na^+ + 4e^-$$

$$4\ Na^+ + 3Na_2S_5 + 4e^- \rightarrow 5\ Na_2S_3$$

In this one-phase region the voltage drops gradually from 2.08 V to 1.78 V. Discharging beyond the composition of Na_2S_3 causes a steep decline in performance because solid sulfides (Na_2S) are being formed. Further details on Na/S cells and the electrochemical reactions can be found elsewhere (1,2).

LiAl/FeS SYSTEM

In contrast to Na/S, where both active materials are liquid, LiAl/FeS cells comprise solid electrodes and a liquid electrolyte at temperatures of about 450°C. The cells are prismatic and contain several positive and negative electrodes (multiple design). In the charged state the negative electrode consists of the alloy LiAl, the positive of FeS. Both electrodes are porous and are soaked with the electrolyte which is LiCl/KCl eutectic (or LiCl rich).

During discharge Li_2S and Fe are formed :

$$2\ LiAl \rightarrow 2\ Al + 2Li^+ + 2e^-$$

$$2\ Li^+ + FeS + 2e^- \rightarrow Li_2S + Fe$$

Thus, in the discharged state the negative electrode consists of a porous Al-matrix and the positive of Li_2S and Fe. However, the reaction is actually more complex than shown; for example, an intermediate compound, $LiK_6\ Fe_{24}\ S_{26}$ Cl (J phase) is formed through an interaction with the KCl in the electrolyte. There is a voltage plateau at 1.33 V (open circuit voltage). The electrodes (LiAl and FeS) have the shape of plates and are separated by a boron nitride felt or a MgO powder separator. Because the cells tend to swell during cycling, they are held in place by restriction plates on the outside. The cells, having capacities between 195 and 395 Ah, are at present operated under cover gas, because the current feedthroughs are not leak tight. Further details on LiAl/FeS cells and the electrochemical reactions in the FeS electrode and in the melt can be found elsewhere (3,4).

DATA AND PROPERTIES OF THE TWO SYSTEMS

The most important properties of rechargeable battery systems for use in electric vehicle traction and load levelling are :

- specific energy of cells and batteries (volumetric and gravimetric)
- maximum specific power of cells and batteries (volumetric and gravimetric)
- lifetime
- battery characteristics
- safety of cells and batteries
- current key technical barriers
- economic considerations

SPECIFIC ENERGY

Specific energy is determined mainly by cell voltage and cell weight. Cell resistance is not important with respect to this property. In Table I the specific energy is listed for single cells and batteries. It can be seen that Na/S shows considerably higher values. Therefore, Na/S can be characterised as a high energy system.

Both systems have an initial utilization of 85 to 90% and a capacity decline of 0.02 to 0.05 % per cycle. At high discharge rates (1h)

Na/S does not lose capacity whereas with LiAl/FeS the utilization drops to about 50% of the value at the lower rates.

TABLE I. SPECIFIC ENERGY FOR SINGLE CELLS AND BATTERIES

Cell Type		Wh/kg	Wh/l
LiAl/FeS			
EPI Group VI	single cells	90	225
Gould P 113	single cells	75	185
LCMP-12	single cells	75	185
Realistic goals	single cells	120	300
Realistic goals batteries		100	150
Na/S			
A.02	single cells	150	270
A.04	single cells	160	290
Realistic goals	single cells	210	380
Realistic goals batteries		150	135

MAXIMUM SPECIFIC POWER

Specific power is determined by cell resistance, cell voltage, and cell weight. Table 2 shows values for single cells and batteries. According to this, present LiAl/FeS cells exhibit a higher power density than Na/S cells. As the power decline during discharge is a factor of about 2 in LiAl/FeS and only about 1.4 in Na/S, the power in the discharged state is almost equal for both systems. However, values for the volumetric power density are clearly in favor of LiAl/FeS. Therefore, LiAl/FeS can be characterised as a high power system.

TABLE 2. SPECIFIC MAXIMUM POWER FOR SINGLE CELLS AND BATTERIES (FULLY CHARGED)

Cell Type		$\frac{W}{kg}$	$\frac{W}{l}$	Fraction of power available in the discharged state
LiAl/FeS				
EPI Group VI	single cells	90	225	0.48
Gould P 113	single cells	175	430	0.23
LCMP-12	single cells	190	470	0.45
Realistic goals	single cells	180	450	
Realistic goals (batteries)		150	225	0.5
Na/S				
A.02	single cells	120	215	0.58
A.04	single cells	140	250	0.64
Realistic goals	single cells	180	320	
Realistic goals (batteries)		145	130	0.7

LIFETIME

LiAl/FeS cells are terminated when the capacity has declined to 80% of its initial value or the coulombic efficiency to 95%. The reasons for failure are mainly feedthrough leakage or extrusion of active material at the edges of the positive electrode. Na/S cells are terminated when the capacity has declined to 70% of its initial value or the cell resistance exceeds 30 mΩ or if there is a short circuit. The main failure mode is the deterioration of the ceramic electrolyte (breaking, cracks, flaws), causing in most cases a resistance decline.

With both systems, lifetimes of more than 1200 cycles could be achieved. However, the average lifetime for LiAl/FeS cells lies between 300 and 450 cycles (4h/8h-rate) and for Na/S around 500 cycles (2h/4h-rate). These figures indicate that Na/S at present may be more advanced especially when the high rates are being considered with which the cells are tested at the present. Another point in favor of Na/S is that operation is already in open air whereas failure in LiAl/FeS cells due to leakages were excluded a priori by operating them still under covergas. However, with both cell types, no principle obstacle could be identified which would prevent longer lifetime.

In the case of LiAl/FeS, 3 modules (each about 4kWh) with Mark II cells have been built and operated so far. Two of them accomplished more than 200 cycles (one was terminated after 100 cycles). With Na/S a few modules (about 1kWh each) could be operated up to 200 cycles. Two batteries were successfully tested in a VW van under driving conditions.

Of major importance for battery operation is the rate of early failures. In order to minimize maintenance all cells in a battery unit should fail at almost the same time. With LiAl/FeS cells the rate of early failures is remarkably low : from two groups of 24 Eagle-Picher cells only one cell failed below 100 cycles and from 26 Gould cells no failure was observed. Na/S cells at present show a higher rate of early failures : within a rather good cell group about 1 out of 10 cells must be expected to fail below 100 cycles.

Therefore, at present Na/S exhibit longer lifetimes, whereas LiAl/FeS has the advantage of a lower rate of early failures.

BATTERY CHARACTERISTICS

LiAl/FeS modules at present consist of a series connection of 10 single cells with about 4 kWh at the 4h discharge rate (Eagle Picher). For an 48 kWh electric vehicle battery 12 modules are connected in series (144 V). At todays low discharge/charge rates no cooling is necessary (temperature rises from 470 to about 500°C). However, at the 2h/4h rate cooling will be necessary. The casing of the module or the battery must contain a cover gas (He or N_2). Therefore, cooling must be done indirectly by tubes. LiAl/FeS cells cannot be overcharged, because beyond 1.7 V the cells become short circuited. Also overdischarge should be avoided, because in this case the polarity of a cell will be reversed. Protection against overcharging is done by "monitoring" each single cell and by "equalizing" : If one cell reaches 1.6 V, the constant current mode is changed into a constant voltage mode where each cell is limited to 1.6 V. This "equalizing" procedure requires some extra wiring and electronic equipment.

Na/S-modules at present consist of 2 parallel connected strings of 8 series connected cells with about 1.8 kWh at the 2h-discharge rate. Todays approach for a 50 kWh electric vehicle battery is the parallel connection of 2 modules and the series connection of 14 such submodules (190 V).

Cooling of the battery can be done either directly by air or by tubes as in LiAl/FeS. High discharge rates can easily be reached because the available range of operating temperature is relatively large (from 300 to 400°C). Charging can be done without equalizing. However, if after some time cells show differences in capacity, some kind of equalizing may be necessary with Na/S as well. No protection against overdischarge is necessary.

SAFETY OF CELLS AND BATTERIES

Safety is a crucial point with a battery which has to operate in an electric vehicle. Both systems have been tested by mechanically destroying a hot cell. With LiAl/FeS cells no hazardous effects could be observed (no fire, no explosion, no gases), whereas with Na/S cells such a treatment resulted in fire and emission of noxious gases (however, no explosion occurred). Squeezing of Na/S cells is possible without leakage of the seal. Another possible safety hazard can arise when the operating temperature is increased accidentally (e.g. by a short circuit). With LiAl/FeS at about 150°C beyond the operating temperature the negative electrode melts and a short circuit will be the result (but this represents no hazard). Na/S can be operated up to about 170°C beyond its operating temperature. However, if a temperature increase of more than 200°C occurs, the seal may not bear the sulfur pressure and thus may leak. Release of noxious gases or fire might be the consequences.

A short circuit of LiAl/FeS generally melts the negative electrode, but poses no hazard; a short circuit of Na/S leaves the cell intact, because the temperature increases only about 100°C. Another point is overcharging : LiAl/FeS cannot be charged beyond 1.7 V, because this would result in a short circuit. With Na/S it is possible to charge up to 30 V. However, beyond this value the ceramic can be destroyed which, in consequence, can give rise to cell leakages. These safety aspects are listed in Table 3. In summary, todays LiAl/FeS cells can be overcharged, overheated, short circuited and mechanically destroyed without any hazard. In contrast, Na/S cells, although safe under normal operating conditions, may be unsafe when overcharged, overheated or destroyed. Therefore, additional safety precautions are necessary.

TABLE 3. SAFETY CHARACTERISTICS

Incident	LiAl/FeS	Na/S
Mechanical destruction of a hot cell	no fire, no explosion, no noxious gases	fire, gases
Overheating	If more than 150°C beyond operating temperature : short circuit	If more than 200°C beyond operating temperature, seal leakage (gases, fire)
Overcharging	If U > 1.7 V : short circuit	Up to about 30 V no hazard, beyond this value : degradation of ceramic (explosion, fire)
Short circuit	No safety hazard	No safety hazard (cells remain sealed)
Direct reaction of the active materials	Li + FeS : only moderate temperature increase (formation of solid Li_2S)	Na + S : violent reaction

Experiments designated to investigate the safety of complete batteries have not been carried out so far. However, with LiAl/FeS the failure of the 20 kWh Mark IA battery showed that there will probably be no safety hazard : although the temperature during the 15-minute total short circuit rose locally to 1000°C, nothing happened. The casings of the cells as well as the battery housing were not affected. However, nothing is known about the mechanical destruction of such a battery, especially, if different separators are used (e.g. MgO).

With Na/S single cells it is known, that a short circuit gives rise to about 100°C increase in temperature. As neighbour cells will not be affected by such a temperature, failure propagation in a battery by this mechanism will not occur. However, destroying a Na/S battery mechanically could result in an explosion type reaction. Therefore, measures must be taken in order to prevent a temperature increase of more than 200°C (providing several barriers, emergency cooling, safe location of battery in car).

Therefore, Na/S batteries will require considerably more safety measures in order to meet possible accidents than LiAl/FeS batteries. However, as there is only little experimental evidence at present, such an assessment can only be preliminary.

CURRENT KEY TECHNICAL BARRIERS

Both systems have areas which need more R + D work in the future. In the following, the most important problems are described. With LiAl/FeS the main problem today is to achieve longer lifetime. A great step in this direction would be the development of a feedthrough which is leaktight and which is not subject to internal short circuits due to creeping electrolyte. Another necessity is to develop cells with MgO-separator instead of the expensive BN-felt. Recently, promising results have been obtained with this type of separator at Gould. However, the power density in the discharged state and the energy density have to be improved further.

In the case of Na/S the predominant task is to reduce the rate of early failures. This is a vital requirement for the operation of large batteries. This involves the optimization of the ceramic and the sealing areas. Further areas of R + D must be the improvement of the cell casing (corrosion resistant, electrically conductive, low cost coatings) and the provision of additional safety features. Common to both systems is the need for a battery case (low weight, low volume, good insulation, low cost). The Linde Division of Union Carbide has manufactured a case (load bearing vacuum insulation) which can be expected to exhibit a heat loss of less than 300 W for a 50 kWh vehicle battery.

ECONOMIC CONSIDERATIONS

At this stage of development only rough estimations can be made on investment and operation costs of these batteries. There are cost analyses carried out by Argonne (5), Ford (6), Marcoussis (7) and Hittman Assoc. (8). They indicate a selling price between 30 and 100 $/kWh depending on the application. Within the accuracy of the estimations, no difference in cost between the two systems could be detected. In the Na/S system the most uncertain item is the cost of the β-Al_2O_3 ceramic, whereas in the LiAl/FeS system the cost of lithium (because it has to be recycled) and of the separator (expensive BN-felt or cheap MgO) are rather unknown factors.

Another point which affects the economic feasibility, is the availability of raw materials. Whereas in the case of Na/S all materials are abundant, lithium is restricted. In addition, it is not known to what extent lithium will be used for a future fusion technology. As the USA have considerable lithium resources of its own, this point will be of minor importance for this country. However, for Germany, the use of lithium will be a considerable drawback, because most of the lithium needed for large quantities of batteries must be imported, (e.g. from Chile or U.S.S.R.).

COMPARISON OF THE Na/S AND THE LiAl/FeS SYSTEM

Based on the collected data, 25 different properties of each system were assessed in Table 4, by giving each of the properties a certain rating (from 1 = "poor" to 4 = "excellent"). Four different scenarios were considered : (A) the most likely one, (B) optimistic for LiAl/FeS, (C) optimistic for Na/S and (D) optimistic for both systems (see Table 5). The ratings were multiplied by an "importance factor" (from 1 = "not very important" to 4 = "vital"). This factor differs according to whether the application in the U.S.A. (electric vehicle and load levelling) or in the F.R.G. (mainly electric vehicle; no Li-resources) is being considered. The sum total of all the individual ratings, each multiplied by its importance factor, reflects directly the degree of qualification of a certain system for a certain application. Table 5 shows the sum total of the ratings. It can be seen that both systems show a comparatively high degree of qualification as high performance batteries. If application in Germany is being considered, it turns out that the Na/S system as a whole will be more suitable than LiAl/FeS. For application in the U.S.A., however, both systems will be equally suitable.

The results of the study presented here are of course preliminary and the assessment may be different if one of the systems makes a larger progress. Therefore, a revision will be necessary in the course of time.

TABLE 4. QUANTITATIVE ASSESSMENT OF PROPERTIES

Property	Na/S		LiAl/FeS	
	Rating ①	Importance ② (Germany) ③	Rating ①	Importance ② (U.S.A.) ④
Gravim. Power (Battery)	3	3	3	2
Volum. Power (Battery)	3	3	4	3
Gravim. Energy (Battery)	4	3	3	2
Volum. Energy (Battery)	4	3	4	3
Operating Temperature	2	2	1	1
Range of Oper. Temp.	3	3	2	2
High Test Rates (Long Term)	3	3	2	2
Rapid Charge/Discharge	3	3	3	2
Capacity Decline	2	2	2	3
Power Decline at Discharge	3	3	2	2
Utilization at High Rates	3	2	2	1
Connection Mode of Cells	2	2	3	2
Rate of Early Failures	2	4	3	4
Cycle Life (Cells)	2	3	2	4
Self Discharge	4	1	2	1
Thermal Cycling	2	3	2	2
WH-Efficiency	2	1	3	3
Sturdiness	2	4	2	2
Safety (Batteries)	2	4	3	4
Battery Atmosphere	2	2	1	1
Overcharge/Overdischarge	3	2	1	1
Probability of Solving Key Problems	2	4	2	4
Availability of Raw Materials	4	4	2	2
Manufacturing Properties	2	3	2	3
Simplicity of Recycling	2	2	1	1

① : 1 = poor
2 = acceptable
3 = favorable
4 = excellent

② : 1 = not very important
2 = important
3 = very important
4 = vital

③ : Germany : mainly for electric vehicles,
Li has to be imported

④ : U.S.A. : electric vehicle + load levelling,
Li readily available

TABLE 5a. FINAL ASSESSMENT USING THE MOST LIKELY SCENARIO

Application in the F.R.G.

Na/S	LiAl/FeS
182	162

Application in the U.S.A.

Na/S	LiAl/FeS
145	141

TABLE 5b. FINAL ASSESSMENT USING AN OPTIMISTIC SCENARIO

Na/S : - only failure mode is short circuit
- rate of early failures reduced
- safety improved

LiAl/FeS : - 500 cycles also with 2h/4h-rate
- feedthrough sealed (operation in air)
- gravimetric energy increased considerably

Application in the F.R.G.

Na/S	LiAl/FeS
192	170

Application in the U.S.A.

Na/S	LiAl/FeS
155	146

REFERENCES

1. W. Fischer : "State of Development and Prospects of Na/S Batteries", Solid State Ionics 3/4 (1981) 413.

2. B. Cleaver and A. J. Davies : "Properties of fused Polysulfides". Electrochimica Acta 18 (1973) 733.

3. D. L. Barney : "Lithium/iron Sulfide Batteries for electric Vehicle Propulsion and Other Applications", Progress report for October 1979 to September 1980, Argonne National Laboratory Report ANL-80-128 (February 1981).

4. Z. Tomczuk, S. K. Preto and M. F. Roche : "Reaction of FeS electrodes in LiAl-KCl electrolyte", J. Electrochem. Soc. 128 (1981) 760.

5. A. A. Chilenskas et al. : "A preliminary estimate of the manufacturing cost for Lithium/Metal Sulfide Cells", ANL-Report ANL-79-59, January 1980.

6. Ford Comp., "Sodium-sulfur battery ecenomic analysis", DOE-Report DOE/CH_10012-TI-Sept. 1980.

7. A. Wicker et al. : "Economic Assessment and Comparison of Alternative Beta Alumina Electrolyte", EPRI-Report EM-1799, April 1981.

8. D. Sullivan et al. : "Life Cycle analysis of electric vehicle storage batteries", Final report, Hittman Assoc., Columbus, Md., Dec. 1980.

II. POLYELECTROLYTES

STATIC AND DYNAMIC PROPERTIES OF POLYELECTROLYTES

F. Nallet, J.P. Cotton, M. Nierlich, G. Jannink
Laboratoire Léon Brillouin
CEN-Saclay 91191 Gif-sur-Yvette, Cedex, France

Results obtained recently from elastic[1),2)] and quasielastic[3),4)] neutron scattering experiments in concentrated polystyrene sulfonates solutions, give us the opportunity to analyze in more detail the static and dynamic structure of polyelectrolytes. The neutron small angle scattering technique has the advantage to allow a better discrimination between different partial structures, and this is particularly usefull in the study of multicomponent systems.

We discuss here the partial static structure and form factors, the counterion self diffusion and the collective longitudinal modes.

I. STATIC PROPERTIES

Static properties of polyelectrolytes made of flexible macromolecules were first derived[5)] from observed dependences of viscosity and osmotic pressure with the concentration. From such observations, it was concluded that, at small concentrations the polyion molecule is stretched as a rod and that both polyion and counterion distributions in space are strongly correlated. Here we consider the observed small angle neutron scattering cross sections in concentrated solutions as a function of wave vector transfer q and of concentration. We wish to determine the size of the polyions, their distribution in space as well as distribution of the counterions. The differential cross section of a polyion solution in water without added salt is :

$$\frac{d\sigma}{d\Omega} = b_1^2\, S_{11}(q) + 2b_1 b_2 S_{12}(q) + b_2^2\, S_{22}(q) \qquad (1)$$

where b_α is the contrast length of a molecule of type α (α=1 for counterions, α=2 for polyions), and where $S_{\alpha\beta}(q)$ are the partial structure factors

$$S_{\alpha\beta}(q) = \sum_{I\in\alpha}\sum_{J\in\beta}\sum_{ij} \langle e^{i\vec{q}.(\vec{r}_{Ii}-\vec{r}_{Jj})}\rangle \qquad (2)$$

(I,J) and (i,j) are respectively the molecule and monomer indices.

Because it is possible to adjust the contrast length b_α of a given species α, to a chosen value, the different partial structure factors $S_{\alpha\beta}(q)$ may be derived separately from the experiment. Figure 1 shows for instance the counterion and the polyion structure factors (S_{11} and S_{22}) for PSS(TMA) in D_2O at a concentration of 16% by weight (590 monomoles m^{-3}).

These structure factor indicate a pronounced maximum. The wave vector q_m at which a maximum occurs increases with concentration C as $C_2^{1/2}$ (where C_2 is the polyion concentration). The interpretation of the S_{22} structure peak in terms of polyion distributions raises fundamental questions. If we consider simple electrolytes, the distribution in space is characterized by the spherical ionic atmosphere surrounding each charge. But in the case of polyelectrolytes, there is large charge dissymetry.

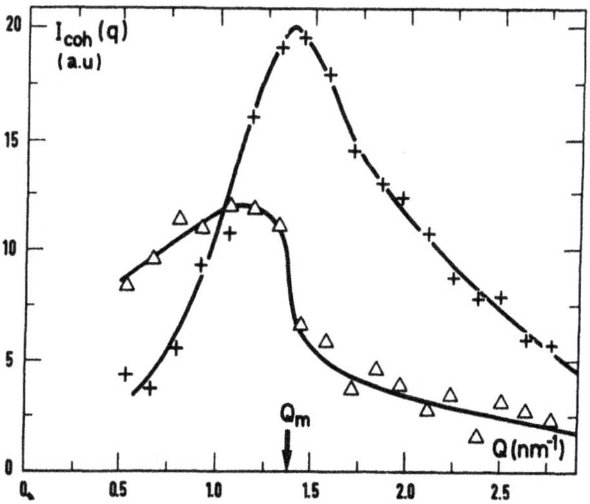

Fig.1 - Differential cross sections as a function of the wave vector q.
▲ PSS_D TMA_H in D_2O : Counterion-Counterion structure factor.
\+ PSS_H TMA_D in D_2O : Polyion - Polyion structure factor.

Each "polyion is the center of its own counterion atmosphe and cannot be assumed to participate in the atomsphere of other ions"[5]

A first interpretation of Fig.1 consists to associate the maximum $S_{22}(q_m)$ with a 2 dimensional order of stretched coils. Further experimental evidence shows however that the polyions are not fully stretched at these high concentrations. The polyions can therefore not align sufficiently for such an ordering. The second interpretation consists in assuming the existence of an entangled system of random coils with large persistence length ("isotropic[6] model"). The observed peak is then a consequence of the long range repulsive Coulomb interaction between the coils. A structure peak may be derived in the random phase approximation (RPA)

$$\frac{1}{S_{22}(q)} = \frac{1}{{}^{o}S_{22}(q)} + \frac{1}{\beta\hat{V}(q)} \quad . \tag{3}$$

${}^{o}S_{22}(q)$ is the polyion-polyion structure factor in absence of Coulomb interaction, and $\tilde{V}(q)$ is the Fourier transform of the screened potential

$$V(r) = \frac{e^2}{\varepsilon r} e^{-H/\xi} \quad . \tag{4}$$

The screening length ξ varies with concentration as

$$\xi = \left(\frac{4\pi e^2 \beta c}{\varepsilon}\right)^{-1/2} \quad . \tag{5}$$

Formula (3) exhibits a maximum, $S_{22}(q_{mRPA})$. It fails however to reproduce satisfactorily the observed wave vector dependence of the polyion-polyion structure factor. It also fails to reproduce the observed concentration dependence of q_m, if, as pointed out by de Gennes[3]), the polyion configuration is characterized by a large persistence length $b \propto C^{-1}$ (for a random coil without persistence length, $q_{mRPA} \propto C^{1/2}$: with the above persistence length, $q_{mRPA} \propto C^{2/3}$). Generally, the RPA is found to be inadequate for all systems having large coupling constants. The coupling constant is here

$$\Gamma = \frac{Q}{a} \propto 1 \qquad (6)$$

where $Q = \beta e^2/\varepsilon$ is the Bjerrum length and where a is the distance between charges.

A third interpretation, which is more pragmatic, consists in writting the structure facture as a sum of two terms

$$S_{22}(q) = {}^{F}S_2(q) + {}^{D}S_{22}(q) \qquad (7)$$

where ${}^{F}S_2$ is the contribution of the diagonal terms I=J in (2) (form factor) and where ${}^{D}S_{22}$ is the contribution of the nondiagonal terms (distinc pair correlations).

We can write further

$$S_{22}(q) = {}^{F}S_2(q)\ G_2(q) \qquad (8)$$

where

$$G_2(q) = 1 + {}^{D}S_{22}(q)/{}^{F}S_2(q) . \qquad (9)$$

Fig.2 shows the wave vector dependence for the function $G_2(q)$, as measured in a system of rigid polyelectrolyte spheres and as assumed for flexible polyelectrolytes

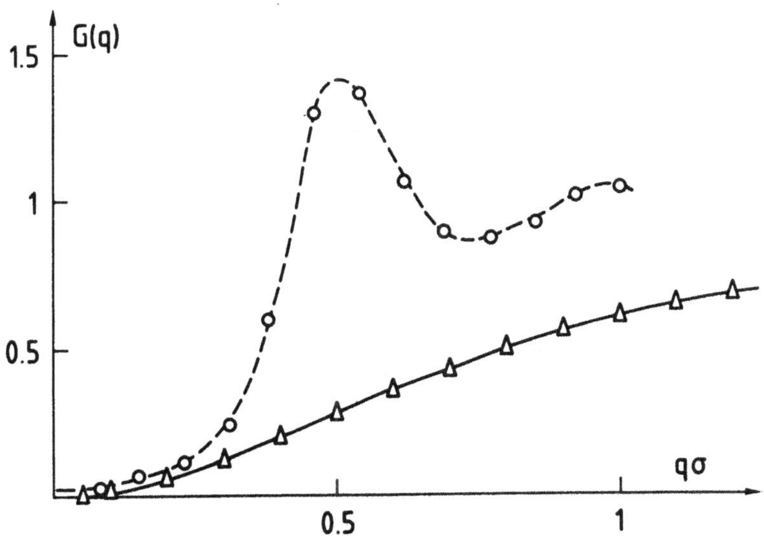

Fig.2 - Wave vector dependence of the function G(q) defined in equation (9).
—— assumed for polyions made of flexible coils.
--- calculated and measured for polyions made of rigid spheres.

Notice that $G_2(q)$ exhibits space ordering in the case of rigid spheres[7]), whereas no such ordering is predicted in the case of concentrated flexible polyions. The low value for $G_2(0)$ reflects the repulsive Coulomb interaction. The effective $G_2(q)$ function for flexible polyions may deviate from the assumed monotomic increase. However no periodicity in the q dependence is expected. Recent experimental results show indeed that this is the case. The function $G_2(q)$ is determined by separate measurements of $S_{22}(q)$ and ${}^{F}S_2(q)$, which is again made possible by use of the deuterated labels[2]).

In the case of the counterions, the structure factor $S_{11}(q)$ may be written

$$S_{11}(q) \cong G_1(q) \tag{10}$$

because the form factor of the small counterion sphere is constant in the observed q range (the molecule has a diameter smaller than 4Å). Figure (1) indicates :

$$S_{11}(q) > S_{22}(q)$$

for $q < .5\ Nm^{-1}$. This reflects the charge dissymetry between counterion and polyion.

We further notice that $S_{11}(q)$ deviates strongly from the R.P.A. formulation (3), which predicts here a monotonously increasing function of q.

The cross correlation structure factor $S_{21}(q)$ was derived from the data of three different scattering experiments, performed at the same concentrations but with different labeling[8]). The result is shown in fig.3

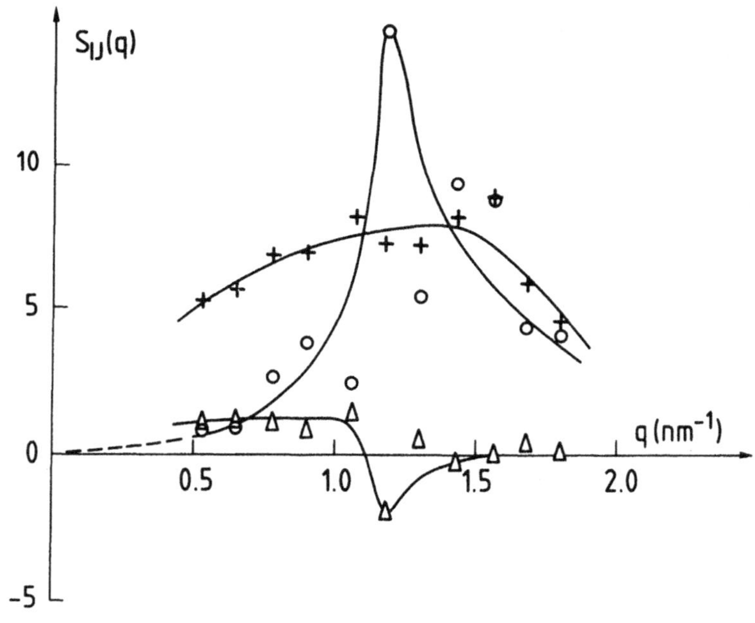

Fig.3 - Partial structure factor $S_{21}(q)$ ▲
Number structure factor $S_{NN}(q)$ +
Charge structure factor $S_{ZZ}(q)$ o

The number structure factor

$$S_{NN}(q) = S_{11}(q) + 2\ S_{21}(q) + S_{22}(q)$$

and the charge structure factor

$$S_{ZZ}(q) = Z_1^2\ S_{11}(q) - 2Z_1Z_2S_{12}(q) + Z_2^2S_{22}(q) \tag{11}$$

were obtained by combining the partial structure factors. The charge ratio $|Z_1/Z_2|$ was assumed to be equal to one. In fact, it obeys the sum rule[3])

$$|Z_1/Z_2| = S_{22}(o)/S_{11}(o) .$$

II. THE COUNTERION SELF DIFFUSION

Results of osmotic pressure measurements in polyelectrolytes against pure water, have suggested two "states" for the couterions[10]) : the "free" counterions whose thermal motion contributes to the osmotic pressure, and the "condensed" counterions, whose thermal motion constrained along or around the polyion, does not contribute to the osmotic pressure. There is a rate of exchange between the charged particles belonging to each "state".

Consider now the self diffusion of the counterions. This is a process characterized asymptotically by the diffusion coefficient

$$D_1 = \lim_{t\to\infty} \frac{<|\vec{r}_1(t) - \vec{r}_1(o)| >}{6t} . \qquad (12)$$

Such a coefficient is best measured using a tracer. A more general characterization of the diffusion process is given by the time dependent diffusion coefficient

$$D_1(t) = \frac{<|\vec{r}_1(t) - \vec{r}_1(o)|^2>}{6t} . \qquad (13)$$

Depending upon the time interval (o,t), there are different appropriate ways to measure this coefficient. For instance, a value

$$D_1(t) = 4.4 \times 10^{-6} \text{ cm}^2 \text{ sec}^{-1}$$

is given by P. Callaghan[11]), as a result of an N.M.R experiment in a pulsed field gradient, such that $t \sim 10^{-3}$sec. (This value corresponds to PSS TMA in water without added salt, at a concentration of 8% by weight, and at a temperature T=25°C).

We have observed self diffusion in a much shorter time span, using the quasi-elastic neutron scattering experiment. The sample is $PSS_D TMA_H$ in D_2O at a concentration of 16% by weight. The neutron scattering by such a sample is mostly incoherent, for wave vectors $q \geq 1$ nm^{-1}. As a result the doubly differential cross section may be written as

$$\frac{\partial^2\sigma}{\partial\Omega\partial\omega} = \frac{\sigma_{inc}}{4\pi} \sum_{i=1}^{\mathbb{N}_1} f_i(q,\omega) \qquad (14)$$

where $f_i(q,\omega)$ is the contribution of a counterion. For a purely Fickian diffusion process, the broadening of the $f_i(q,\omega)$ is Lorentzian

$$f_i(q,\omega) = \frac{D_1 q^2}{(D_1 q^2)^2 + \omega^2} . \qquad (15)$$

Counterions in a polyelectrolyte solution are not expected to diffuse according to the idealized Fick law. Since however the experiment is performed in a limited range

$$\left.\begin{array}{l} 7 \times 10^{-1} \text{Nm}^{-1} \leq q \leq 3 \text{ Nm}^{-1} \\ 10^{-11} \text{sec}^{-1} \leq t \leq 10^{-9} \text{sec}^{-1} \end{array}\right\} \qquad (16)$$

we have used the Lorentzian form to interpolate the data. The result is :

$$D_1 = 3.7 \times 10^{-6} cm^2 \ sec^{-1}$$

at room temperature.

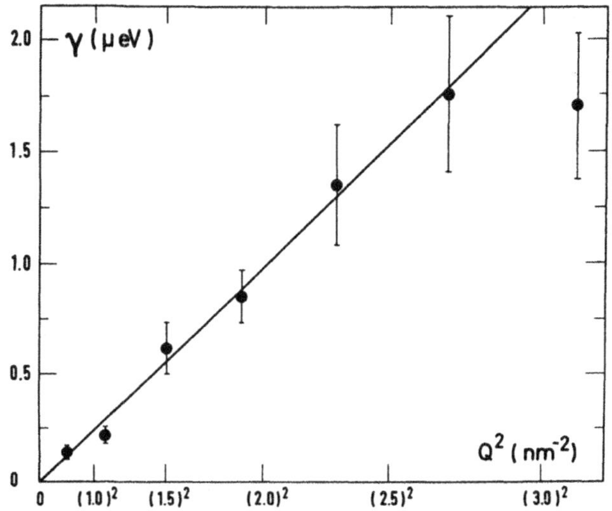

Fig.4 - Half width at half maximum of the doubly differential scattering cross section broadening, as a function of the wave vector q. The sample is PSS_DTMA_H in D_2O at room temperature, at a concentration of 16% by weight.

Fig.4 indicates that the broadening is proportional to q^2 in the interval (16). If the process were Fickian, this result would hold true in larger time and space intervals. We interpret our data as follows.

1) We may assume, with Manning[12]), that the counterion self diffusion coefficient D_1 reflects both a diffusive motion due to concentration gradients, and a convective motion due to electric field gradients. The convective motion results here in a slowing down of the diffusive process, and as a result

$$D_1 = D_{10} \times f$$

where D_{10} is the diffusion coefficient in absence of convection, and where f is a reduction coefficient. Data derived elsewhere on D_{10} and f agree with our result.

2) The self diffusion coefficient $D_1(t)$ is predicted to have a pronounced time dependence, because the velocity correlation function obtained from computer simulation decays as slowly[13]) as $t^{-S/2}$, instead of the exponential decay associated to a Fickian process. The time dependence for $D_1(t)$ is not observed in our experiment because its characteristic time is too large compared to our observation time interval.

3) Deviations from the Fickian process due to exchange between the two sorts of counterions (condensed and free) do not show up because the rate of such exchange lies outside of our time interval.

Measurement of the broadening over a larger reciprocal space interval would probably indicate deviations from the Lorentzian form.

III. THE COLLECTIVE LONGITUDINAL MODES. THE SLOW CONCENTRATION DIFFUSION COEFFICIENT

The polyions and counterions dispersed in water form a strongly coupled two component system. As a result, the time decay for the correlations in concentrations of either species $\alpha = 1,2$

$$<\delta\tilde{C}_\alpha(\vec{q},t) \quad \delta\tilde{C}_\beta(-\vec{q},t)>$$

may be represented by the superposition of two relaxation processes, called the fast and slow relaxations. If we measure the time dependent differential cross section

$$F(\vec{q},t)$$

in a time interval $[0,t_1]$ large compared to the fast relaxation time $T_F(q)$, the observed time dependence mostly reflects the slow relaxation process.

$$F(\vec{q},t) \simeq S(q)\ e^{-t/T_s(q)} \tag{17}$$

where $S(q)$ is the dynamical structure factor associated with the slow relaxation time $T_s(q)$. We have interpreted quasi-elastic neutron scattering data using formula (17). A a slow collective diffusion coefficient was then derived from the wave vector dependence of this relaxation time

$$D_s(q) = 1/(q^2\ T_s(q)) \ . \tag{18}$$

In polyelectrolytes without added salt, $D_s(q)$ has a pronounced q dependence. When however salt is added, $D_s(q)$ is constant. A general formulation for the concentration diffusion coefficient in a one component system (which we label here with the index 2) is

$$D(q) = \frac{\mu_2(q)}{\beta S_{22}(q)} \tag{19}$$

where $\mu_2(q)$ is the mobility and $S_{22}(q)$ the structure factor. For the interpretation of the scattering experiment, we have identified $D_s(q)$ and $D(q)$. Figure 5 shows $D_s(q)$ for a concentration of 16% by weight. The product $D_s(q)\times S_{22}(q)$ is seen here to be constant in the interval $[0,q_m]$, and to decrease with q, as $q \geq q_m$.

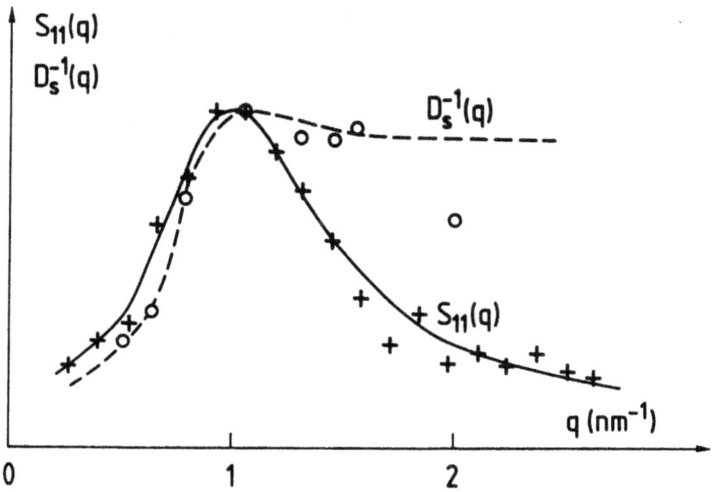

Fig.5 - Variations of inverse concentration diffusion coefficient $D_s^{-1}(q)$ with wave vector q. The polyion structure factor $S_{22}(q)$ is represented on the same scale.

Theory[3]) predicts that $\mu_2(q)$ decreases with q in the same manner as the polyion form $^{F}S_2(q)$, for $q > b^{-1}$, where b is the persistence length. Note that similar scattering experiments, in which however the polyions are made of rigid molecules, are interpreted with the formula

$$D(q) = \frac{\mu_2(o)}{\beta G_2(q)} \tag{20}$$

where $G_2(q) = 1 + \frac{^{D}S_{22}(q)}{^{F}S_2(q)}$.

The formula (19) and (20) are similar in the domain of reciprocal space when both types of polyions are considered as rigid.

A prediction is given in figure 6 for the concentration and wave vector dependences of the concentration diffusion coefficient, $D_s(q,C)$.

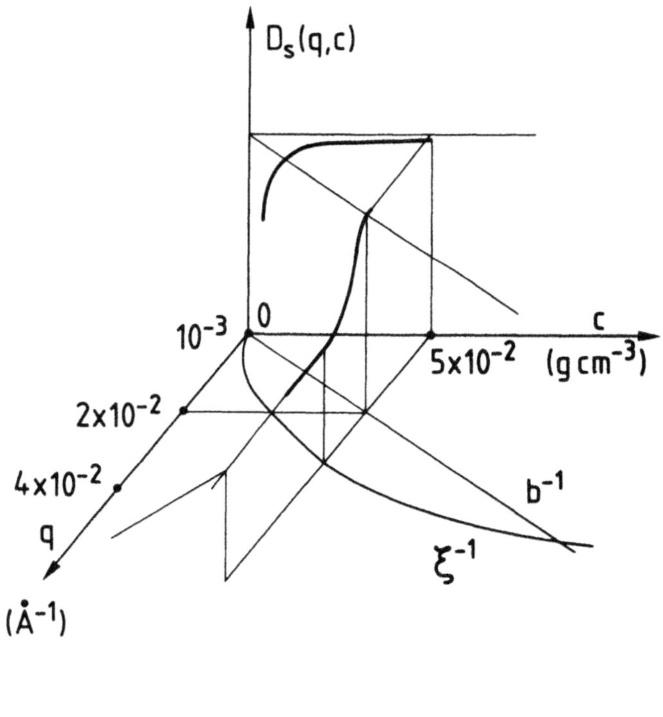

Fig.6 - Predicted concentration and wave vector dependence of the concentration diffusion coefficient $D_s(q)$. The dark line segment parallel to the C axis represents schematically the light scattering results in ref.[14]). The dark line segment parallel to the q axis represents schematically the neutron scattering results in refs.[3])[4]).

It is based on the assumed exsitence of a persistence length $b \propto \xi^2$ and such that $b > \xi$.

o

o o

In conclusion, it may be said that relations are now found between the different observations on static and dynamic structure of polyelectrolytes. A quantity which seemes crucial in holding together formulations of intra- and intermolecular correlations, and which reflects the large charge dissymetry between polyion and counterion, is the polyion persistence length[15]). Its value should be known accurately in order to test the theory. This can in principle be achieved to day from a scattering experiment, because of recent improvements in such techniques.

As for the partial structure factors, they may contribute to determine the "dielectric" structure of polyelectrolytes.

REFERENCES

1) M. Nierlich et al., J. Physique 40, 701 (1979)
2) C. Williams et al., J. Polym. Sc. (Letters) 17, 379 (1979)
3) J. Hayter, G. Jannink, F. Brochard, P.G. de Gennes, J. Physique 41, L451 (1980)
4) F. Nallet, G. Jannink, J. Hayter, R. Oberthür, C. Picot, Submitted to J. Physique
5) A. Katchalsky, "Polyelectrolytes"
6) P.G. de Gennes, P. Pincus, R.M. Velasco, F. Brochard, J. Physique 37, 1461 (1976)
7) J. Hayter, J. Perifold, J. Chem. Soc. Faraday Trans. 77, 1851 (1981)
8) Such experiments were first performed by R. Oberthür on DNA solutions (Mainz, ILL Grenoble)
9) L. Belloni, Thesis, Paris 1982
10) F. Oosawa, "Polyelectrolytes", Marcel Deckker Ed. New York 1971
11) G. Weill, private communication
12) G. Manning, J. Chem. Phys. 51, 934 (1969)
13) W. Hess, R. Klein, Physica 105, 552 (1981)
14) F. Grüner, W.P. Lehmann, H. Fahlbush, R. Weber, J. of Phys. A14, 307 (1981)
15) R. Odÿk, Macromolecules 12, 688 (1979)

THEORY OF POLYELECTROLYTE CONFORMATIONS

Theo Odijk, Dept. of Phys.Chem., Gorlaeus labs, P.O.Box 9502, 2300RA Leiden, Netherlands.

1. INTRODUCTION

In this review we shall consider only linear polyelectrolytes, i.e. linear, flexible polymers in solution which bear a relatively large number of identically charged groups. The solution as a whole is electrically neutral so there are an equivalent number of small ions present. A simple electrolyte like NaCl may be added as desired and in fact it usually is.

One can identify two main areas of research theorists have embarked upon:
a) the interactions of the small ions with the polyions;
b) the statistical physics of the conformations (synonymous with configurations) of the polyelectrolytes. Naturally, these two subjects are not independent. Imagine a fixed configuration of a polyion. It induces a certain statistical distribution of small ions. The ion distribution in turn influences the potential of mean force between parts of the chain, hence the probability of the fixed configuration.

The statistical mechanics of the whole problem - polyion and small ions - at once is much too laborious[1)], so one tends to unravel the problem. The electrostatic forces between various parts of the chain stiffen the chain locally. Thus the coil is rodlike (in some average sense) on a certain length scale determined by the balance between electrostatic and thermal energy. By studying the arrangement of small ions around rodlike polyions, we can obtain a realistic albeit approximate picture of the small ion distribution. This information unables us to set up the free energy pertaining to a certain conformation. The whole procedure must be self-consistent i.e. the chain statistics now ought to lead to an "average conformation" which is rodlike on a long enough length scale.

Refs.(2-9) present several of the key ideas on the distribution of small ions around rodlike polyelectrolytes. Fixman[8)], in particular, has attempted to justify the use of the Poisson-Boltzmann equation[2,4,7)]. There is indeed more justification for its employment compared to that in the simple electrolyte case because of the very high field exerted by a polyion. However, it is an approximate equation at best. Moreover, neither exact solutions nor uniformly valid approximations to it are known for a number of very important cases, thus precluding the formulation of transparent analytical theories.

An attractive though more approximate alternative has been advanced by Oosawa[3)] and Manning[5)] who argue the following. First, the high field in the vicinity of the polyion forces a number of counterions to cluster around the rod. Secondly, the other small ions will be influenced by a much smaller potential because of the lower effective charge so one might assume the Debye-Hückel approximation to be reasonable.

The effective charge plus Debye-Hückel approach to the potential seems to be good enough to obtain almost quantitative insight in the configurational properties of polyelectrolytes. We emphasize its approximate nature - the resulting theory is a reasonable zero order one.

Thus, we take

$$V(\vec{r}_i-\vec{r}_j) = \left(\frac{f^2q^2}{D}\right) \frac{e^{-\kappa|\vec{r}_i-\vec{r}_j|}}{|\vec{r}_i-\vec{r}_j|} \tag{1.1}$$

as the potential of mean force between two elementary charges i and j separated by the vector distance $|\vec{r}_i-\vec{r}_j|$. Here q = elementary charge, D = dielectric permittivity of the solvent, f = effective charge density parameter ≤ 1, κ^{-1} = Debye screening length due to the small ions in solution, $\kappa^2 = 8\pi Qc$ in excess salt conditions with c the concentration of added 1-1 electrolyte and the Bjerrum length $Q = q^2/Dk_BT = 7.14$ A for H_2O at 298 K (T=temperature).

In the Oosawa-Manning picture[3,5)]

$$\begin{aligned} f &= 1 \quad \text{for } A/Q > 1 \\ f &= A/Q \quad \text{for } A/Q < 1 \end{aligned} \tag{1.2}$$

i.e. counterions accumulate in a thin shell surrounding the polyion to renormalize the charge density to a lower value provided the original charge density is high enough.

2. WORMLIKE CHAIN MODEL

Any useful model explaining the configurational properties of polyelectrolytes should satisfy the following requirements:

a) It must relate to rodlike models which work so well in elucidating various phenomena that are more or less independent of conformation[2)].

b) At extremely low ionic strengths - when electrostatic screening by the small ions is minimal - one expects a highly charged polyion to become rodlike, even on a global scale, whereas at high concentrations the coil should be rather pliant although exhibiting an excluded volume effect.

c) It should stress any "universal" features if these exist.

The wormlike chain, a continuous model in which all structural features are smoothed out, has this versatility. Real chains have a discrete character but replacing them by continuous coils can be shown to be reasonable[10)]. The original (uncharged) wormlike chain[11-13)] has only one intrinsic parameter, the persistence length. Nowadays, various extensions are available[14-17)], but we shall focus attention especially on the charged equivalent[18,19)] of the original model.

Now consider the polyelectrolyte as a uniformily charged space curve denoted by $\vec{r}(s)$, where s is the contour distance from one end to that point. The unit vector $\vec{u}(s)$

$$\vec{u}(s) = \frac{\partial \vec{r}}{\partial s} \tag{2.1}$$

with

$$\vec{u}^2(s) = 1 \tag{2.2}$$

is the canonical variable needed in describing the configurational statistics. A total free energy of curvature $\Delta F_t[\vec{u}(s)]$ is associated with every $\vec{u}(s)$ curve[18-20)]

$$\Delta F_t/k_BT = (\Delta F_o + \Delta F_e)/k_BT = \frac{1}{2}L_p \int_0^{\ell} ds \left(\frac{\partial \vec{u}}{\partial s}\right)^2 + \frac{f^2Q}{A^2} \int_0^{\ell} ds \int_s^{\ell} dt \left[\frac{\exp[-\kappa|\vec{r}(t) - \vec{r}(s)|]}{|\vec{r}(t) - \vec{r}(s)|} - \frac{\exp - \kappa(t-s)}{(t-s)}\right] \tag{2.3}$$

where ΔF_o = intrinsic bending energy, ΔF_e = electrostatic free energy of curvature, the polyion has a linear charge density of one elementary charge per A unit of length, ℓ is the contour length of the macromolecule and L_p is the bare persistence length. The first term on the right-hand side of Eq.(2.3) derives from the curvature energy of an elastic rod, its bending force constant being linearly related to L_p[12)]. Torsional terms[17)] can be added if one wants to improve the model but sufficiently general computations tend to become extremely complicated by way of the condition expressed by Eq.(2.2). The second term on the right hand side of Eq.(2.3) is a superposition of potentials like Eq.(1.1). If one assumes a discrete charge distribution with charge spacing A along the curve, the same expression will be obtained whenever $A\kappa \ll 1$.

The statistical average of some property $A[\vec{u}(s)]$ is given by a path integral of the kind[18)]

$$\langle A[\vec{u}(s)]\rangle = Z^{-1} \int_{\vec{u}^2(s)=1} D[\vec{u}(s)]\ A[\vec{u}(s)] \exp -\Delta F_t/k_BT \tag{2.4}$$

with

$$Z = \int_{\vec{u}^2(s)=1} D[\vec{u}(s)] \quad \exp -\Delta F_t/k_BT$$

This represents a nontrivial many-body problem which can however be simplified into two parts[21)]. To begin with one knows that the electrostatic interactions stiffen the chain so it is plausible to postulate that the coil is rodlike on some scale λ (on the average). Next, one assumes $\kappa\lambda \gg 1$, a condition which will be borne out at the end. Hence, for $(t-s)\simeq\lambda$, $\Delta F_e \simeq 0$, so it is expedient to consider two types of interactions short range $(t-s)<\lambda$ and long range $(t-s)>\lambda$. The former cause local stiffening, torsional effects etc., the latter the excluded-volume effect[21,22].

Near the rod limit, expressions of the kind eq.(2.4) simplify considerably so the short-range interaction can be accounted for to a good approximation[18]. The uncharged wormlike chain has the following property[11).

$$\langle\vec{u}(s).\vec{u}(t)\rangle = e^{-(t-s)/L_p} \tag{2.5}$$

which is modified to

$$\langle\vec{u}(s).\vec{u}(t)\rangle \simeq e^{-(t-s)/L_t} \tag{2.6}$$

in the charged case, where the total persistence length

$$L_t = L_p + L_e \tag{2.7}$$

and the electrostatic persistence length is given by[18,20]

$$L_e = \frac{f^2Q}{4A^2\kappa^2} \tag{2.8}$$

Accordingly, from eq.(2.6), we find that the polyelectrolyte is rodlike, on the average, if $(t-s) \lesssim \frac{1}{2}L_t$ whereas no directional correlation is expected when $(t-s) \gg L_t$. Evidently, the length scale λ corresponds to L_t.

From a physical point of view[23], it appears that eq.(2.8) should be accurate if $\kappa L_t \gg 1$. Nevertheless, work still needs to be done to assess the precise nature of what is called the local stiffness approximation[19,23].

A simple dimensional analysis yields the parameter dependence in eq.(2.8).

a) L_e is some function $g(fq,D,T,\kappa,A)$ independent of L_p, λ and ℓ.

b) L_e is proportional to a bending energy $L_e = f^2q^2h(D,T,\kappa,A)$.

c) L_e is independent of discrete effects as long as $A\kappa \ll 1$ and should depend only on the linear charge density (fq/A)

$(L_e = (f^2q^2/A^2)\ j(D,T,\kappa)$

d) κ is a reciprocal length scale, whereas q^2, D and T only match as $q^2/Dk_3T \equiv Q$.

Hence,

$$L_e \simeq \left(\frac{f^2q^2}{A^2}\right)\left(\frac{1}{DT\kappa^2}\right)$$

3. FURTHER DEVELOPMENTS

a. Deviations from Rodlike Behavior

Consider some point 0 of the wormlike polyelectrolyte, together with the unit vector $\vec{u}(0)$, fixed in space. Let the chain take on all possible configurations, each one weighted by a Boltzmann factor. We want to have a look at the properties of the average potential $\langle\Psi(\rho)\rangle$ at point P in a plane perpendicular to $\vec{u}(0)$, the distance OP being ρ.

If the polyelectrolytes were infinitely stiff ($L_p\to\infty$), $\langle\Psi(\rho)\rangle$ would just be a superposition of Debye-Hückel potentials from a very long line charge[5].

$$\langle\Psi(\rho)\rangle = \left(\frac{2qf}{AD}\right) K_o(\kappa\rho) \tag{3.1}$$

where K_o is the zero-order modified Bessel function of the second kind. Chain flexibility causes an increase in charge density along the $\vec{u}(0)$ axis as well as an effective decrease in the average distances from parts of the chain to P. An exact perturbation theory shows that[24]

$$\langle\Psi(\rho)\rangle = \frac{2qf}{AD}\left[K_0(\kappa\rho) + \left(\frac{1}{3\kappa L_t} + \frac{\rho}{3L_t}\right) e^{-\kappa\rho} + \ldots\right] \tag{3.2}$$

with the asymptotic behavior

$$\langle\Psi(\rho)\rangle_{\kappa\rho\to} \sim \frac{2qf}{AD}\left[-\ln\kappa\rho + \frac{1}{3\kappa L_t}\right] \tag{3.3}$$

For highly charged polyions with $A<Q$, $\kappa L_t>1$ if $L_p>Q$, and in fact $\kappa L_t>>1$ usually. Hence fluctuations influence that mean potential at a distance κ^{-1} by only 50% at most. Furthermore, at small $\kappa\rho$, the constant contribution to the potential arising from flexibility is entirely negligible. Thus the effective charge is not modified. Nevertheless, this does not imply that the influence of chain flexibility will never show up in experiments. Activity coefficients should show significant downward shifts[24] as compared to those in theories where rodlike polyelectrolyte models[2-6] are used. However, some compensation will occur because of discrete charge effects[25].

Schurr and Allison[26] have shown that no extra accumulation of counterions occurs due to bending within the context of Manning's reinterpretation[27] of his original[5] counterion-condensation theory. This, of course, corresponds to Odijk and Mandel's viewpoint[24] but the use of the inconsistent model of ref.(27) does not necessarily constitute an improvement.

b. Higher Order Terms

As mentioned earlier, continuous models more complicated than the primitive wormlike chain have been dealt with. Bailey[17], for instance, has attempted to analyze the statistics of helical wormlike chains having a free energy given by

$$\Delta F_h/k_BT = \tfrac{1}{2} L_p \int_0^{\ell} ds \left(\frac{\partial\vec{u}}{\partial s}\right)^2 + \gamma \int_0^{\ell} ds \left(\frac{\partial^2\vec{u}}{\partial s^2}\right)^2 \tag{3.4}$$

where γ is a torsion force constant. Because of the difficult restraint eq.(22), he added a term $\alpha\int_0^{\ell}\vec{u}^2\,ds$ which considerably simplifies the analysis at the expense of obtaining an approximate treatment - a device due to Freed.[28]

Bailey shows that the Kuhn length (equal to 2 L_p for the primitive wormlike chain, and effectively the step length of the equivalent Gaussian coil when the contour length is very long) would increase by a relative amount of order γL_p^{-3}. For polyelectrolytes we can derive the scaling formula for γ as follows

a) γ is some function of (fq, A, D, T, κ).

b) It is proportional to an energy $\gamma = f^2 q^2 h(A,D,T,\kappa)$.

c) If $A\kappa \ll 1$, it must be dependent on the linear charge density. $\gamma = f^2 q^2 A^{-2}\ g(D,T,\kappa)$

d) Finally, γ has dimension $(\text{length})^3$, thus

$$\gamma \cong \frac{f^2 q^2}{A^2 D T \kappa^4} \tag{3.5}$$

The numerical coefficient is in fact 2/5[17]. Thus, one might expect significant corrections of order $f^2 Q/A^2 \kappa^4 L_t^3$ in the polyelectrolyte Kuhn length when torsional terms are included.

However, we must realize that the whole approximation scheme is quantitative only if κL_t is quite large i.e. when the torsional corrections, for example are very small. (See the derivation in ref. 18 and 19). When κL_t is of order unity, the whole procedure of replacing a complicated functional like eq. (2.3) by simpler ones as in eq. (3.4) must be very carefully assessed. This has not been done and further work is sorely needed.

c. Poisson-Boltzmann Results

Marc le Bret has numerically calculated the electrostatic persistence length of a uniformly charged cylinder[29]. The latter is assumed to have a uniform dielectric permittivity different from the solvent. (Le Bret took DNA as a typical example). If one wants to compare the Debye-Hückel approximation with le Bret's results, some care is required. At very low ionic strengths, when the Debye radius is say on the order of 100Å, a curved DNA cylinder is effectively a bent line charge. At an ionic strength of 1M, the Debye screening length is 3Å so from a physical point of view, only local charge interactions are involved.

Accordingly we propose to split up the Poisson-Boltzman results into two parts:

$$L_{e,PB} = L_{e,1} + L_{e,2} \tag{3.6}$$

The parameter $L_{e,1}$ is model dependent whereas $L_{e,2}$ is the contribution one would obtain when viewing the curved cylinder as a line charge. At low ionic strengths $L_{e,1}$ must tend to a constant. At an ionic strength of 0.1M, $L_{e,2}$ is much smaller than $L_{e,1}$ because $L_{e,1}$ is on the order of 3Å ($L_e \simeq 3$Å from eqs. (2.8) and (1.2)) and $L_{e,PB}$ is 93Å from le Bret's paper. Furthermore, $L_{e,1}$ should not increase substantially when the ionic strength is lower than 0.1M, for the Debye radius at 0.1M is already on the order of the DNA cylinder radius. In this way we obtain Table I.

c(M)	$L_{e,PB}$(Å)	$L_{e,1}$(Å)	$L_{e,2}$(Å)	L_e(Å)	f_{PB}/f
0.0001	2300	90	2210	3240	0.83
0.0001	489	90	399	324	1.11
0.0005	222	90	132	64	1.44
0.01	168	90	78	32	1.56
0.02	137	90	47	16	1.71
0.1	93	90	-	3.3	-
0.2	79	79	-	1.6	-
0.6	55	55	-	-	-
1.0	42	42	-	-	-

Table I. Theoretical results of the electrostatic persistence length.

In Table I, the quantity $L_{e,PB}$ has been calculated by le Bret[29] who assumed that the dielectric constant within the cylinder is only one tenth that of H_2O at 298K, i.e. 7.85. The charge density ratio $f_{PB}/f \equiv (L_{e,2}/L_e)^{\frac{1}{2}}$ signifies the relative change in the effective charge density in comparison with the constant value of eq.(1.2). For DNA, we have A 1.7Å so f=A/Q=0.24. The charge density ratio does not deviate very much from unity. The reason is that compensation occurs. Near the cylinder the Manning-Oosawa approach underestimates the effective charge density whereas at distances rather larger than the Debye-Hückel length, the Poisson-Boltzmann potential is lower than that of Manning and Oosawa[8]. Fig.1 of Fixman's paper[8] is a useful guide. For instance, at $c=10^{-4}$M, Fixman's $X(\infty)$ representing the deviation from the Manning potential[5] is substantially lower than unity, ultimately forcing f_{PB}/f to be the same.

d. Measurements

Although several studies exist on the extraction of the total persistence length of flexible polyelectrolytes from data like intrinsic viscosity, they are virtually useless because excluded-volume and other effects have been neglected. Much more serious effort has been devoted to deducing the ionic strength dependence of supposedly double-stranded DNA, a polyelectrolyte which can at least be well approximated by a wormlike chain[30]. An impressive array of techniques such as flow birefringence[31], flow dichroism[32], transient electric birefringence[33], light-scattering[34-37], magnetic birefringence[38,39] and time-dependent fluorescence depolarization[40] has not however yielded an unequivocal characterization of this salt-concentration dependence. Several of the studies appear to bear out eq.(2.8) whereas the others do not. It must be noted that none of the experimental techniques are devoid of difficulties concerning the interpretation of data. To do justice to each of them would entail a review paper all by itself. The author hopes to write such a survey in the near future. (See, in this respect, the comments in ref.26)

4. EXCLUDED-VOLUME EFFECT

a. Theory

We have already seen that within the Debye-Hückel approximation $L_t > \kappa^{-1}$ for all κ if $L_p > Q$. Physically this implies that we have local cylindrical symmetry - an ion sees essentially a rodlike piece of the polyion at all concentrations. The original model of the polyelectrolyte with bare persistence length L_p and electrostatic persistence length L_e can be replaced by a wormlike chain with persistence length L_t surrounded by an impenetrable shell of effective diameter d_{eff}.[21] Strictly speaking this holds only if $\kappa L_t \gg 1$. "Short range" electrostatic interactions (short means smaller than or equal to $O(L_t)$ along the contour) are incorporated within L_t, "long-range" or excluded-volume interactions within d_{eff}. This is precisely the analogous separation occurring in the postulate used in formulating the excluded-volume theory of uncharged polymers[12]. There, short-range interactions are lumped together in the Kuhn length A_K, long-range effects being represented by the excluded-volume β_p between Kuhn segments.

An essential difference with polyelectrolyte theory is that $A_K (\equiv 2L_t$ for wormlike chains) and $\beta_{el} \equiv 2\pi d_{eff} L_t^2$ have a definite physical interpretation whereas β_p is as yet an a priori unknown parameter. The form $\beta_{el} \sim d_{eff} L_t^2$ can be derived as follows. Take two pencils with their orientations fixed and consider the excluded-volume between the pencils. It scales as $(\text{length})^2$ times the diameter of the pencils. Next integrate over all orientations which gives solely a numerical factor. The exact calculation is due to Zimm[41] and Onsager[42].

In excluded-volume theory[12], the chain is expanded because of the long-range interaction (if we assume repulsion dominates). The radius of gyration can be expressed by

$$<S^2> = \alpha_s^2 (Z) <S^2>_\theta \qquad (4.1)$$

where α is a function solely of the excluded-volume parameter $Z \equiv (3/2\pi A_K^2)^{3/2} \beta_p (\ell/A_K)^{\frac{1}{2}}$ with ℓ the contour length. The number of persistence lengths $2\ell/A_K$ should be larger than approximately 100 in order that eq.(4.1) holds[43]. The expansion factor $\alpha_s(Z)$ has the following properties

$$\alpha_s^2(Z) \sim 1 + 1.276\, Z - 2.082\, Z^2 = \ldots \qquad Z \to 0 \qquad (4.2)$$

See ref.12. This series is probably asymptotic[44,45], with zero radius of convergence.

$$\lim_{Z\to\infty} \alpha_s^2 \sim \omega\, Z^{2/5} \qquad (4.3)$$

The factor ω is a constant of order unity whereas the exponent 2/5 has, in spite of great efforts, not yet been firmly derived[46-48]. Very many approximate expressions

in the intermediate range exist, whichever one chooses being a matter of taste.
We are now in a position to discuss polyelectrolyte excluded-volume theory[21,22].
The mean-square radius of gyration is given by

$$<S^2> = \frac{1}{3}\, \ell\, L_t\, \alpha_{s,el}^2\, (Z_{el}) \tag{4.4}$$

The factor $\frac{1}{3}\ell L_t$ is the expansion due to the afore mentioned "short-range" interaction and equals the radius of gyration in the hypothetical theta state. Next we switch on the electrostattic excluded-volume effect which leads us to an expansion $\alpha_{s,el}(Z_{el})$. The electrostatic excluded-volume parameter is expressed as

$$Z_{el} = \left(\frac{3^{3/2}}{32\pi^{3/2}}\right) \beta_{el}\, \ell^{1/2}\, L_t^{-7/2} \tag{4.5}$$

The electostatic excluded-volume between the Kuhn cylinders each of length $2L_t$, is written as

$$\beta_{el} = 2\pi\, L_t^2\, d_{eff} \tag{4.6}$$

Within the Debye-Hückel approximation, in combination with the condensation hypothesis, d_{eff} for polyelectrolytes with A<Q is given by[22]

$$d_{eff} \simeq \kappa^{-1}(-\ln Q\kappa + \ln 4\pi + \gamma - \tfrac{1}{2}) \tag{4.7}$$

Here γ is Euler's constant $\simeq 0.57722$.
A scaling argument yields eq.(4.7). A piece of chain of length η (specified below) and charge $q\eta\, Q^{-1}$ may approach another section of the polyelectrolyte which has a potential $\simeq 2q/QD\; K_0(\kappa\rho)$. It is effectively blocked at a distance $\frac{1}{2}d_{eff}$ where the free energy is about k_BT.

$$\left(\frac{q\eta}{Q}\right)\left(\frac{2q}{QD}\right) K_0\, (\tfrac{1}{2}\kappa\, d_{eff}) \simeq k_BT$$

Using the leading asymptotic behavior for $K_0(t) \simeq (\pi/2t)^{\frac{1}{2}} e^{-t}$ we find

$$\frac{1}{\kappa\, d_{eff}}\, e^{-\kappa d_{eff}} = o\, (\frac{Q}{\eta}) .$$

However, η must be of order κ^{-1} since only at this distance and smaller should there be significant contributions. Hence, $d_{eff} \simeq \kappa^{-1}\, [-\ln Q\kappa]$.
Eqs.(4.4)-(4.7) yield

$$<S^2> = \frac{1}{3}\ell L_t\, \alpha_{s,el}^2\left[\left(\frac{3^{3/2}}{16\pi^{\frac{1}{2}}}\right)\ell^{\frac{1}{2}}\, L_t^{-3/2}\, \kappa^{-1}\, \{-\ln Q\kappa + \ln 4\pi + \gamma - \tfrac{1}{2}\}\right] \tag{4.8}$$

The expansion factor $\alpha_{s,el}$ is a monotonically increasing function of Z_{el}, having a maximum at κ_m^{-1} given by

$$\kappa_m^{-1} = O(Q^{\frac{1}{2}} L_p^{\frac{1}{2}})$$

Thus eq.(.48) is very complicated because L_t increases with decreasing salt concentration. Still, $\langle S^2 \rangle$ must increase with decreasing ionic strength because the repulsive force increases. Vice versa, a bound on the behavior of α_s is found. (See ref.(4.9)). If we assume the excluded-volume effect to be large and if we furthermore neglect all numerical coefficients as well as logarithmic corrections, we derive from eqs.(4.3) and (4.8)

$$\langle S^2 \rangle^{\frac{1}{2}} \simeq \ell^{3/5} L_t^{1/5} \kappa^{-1/5} \qquad (4.9)$$

an expression useful in scaling theory[23], in which interpenetrating chains are considered.

b. Comparison with Experiments

Most of the published measurements are of little use. In most papers in which the radius of gyration is measured by light scattering, the concentrations are too high, the effect of polydispersity is unknown, A_2 is not zero for any salt concentration thus invalidating the model, L_p is not known accurately, or the measured radius of gyration simply has too large an error. The author feels that extraction of meaningful information from the experiments for the construction of theories[50] is not fruitful at present.

One paper which contains a useful set of measurements on potassium (poly)styrenesulphonate is that of Raziel and Eisenberg[51]. They assert that $M_w/M_n \leq 1.07$ (M = molecular weight of the polyelectrolyte). One of their samples had a contour length of 4300 Å. From $\langle S^2 \rangle^{\frac{1}{2}} = 120$ Å we derive the bare persistence length $L_p \simeq 10.0$ Å.

Table II shows the ratio of the asymptotic radius (eq.(4.9)) to $\langle S^2 \rangle^{\frac{1}{2}}_{exp}$. If one acknowledges that $\langle S^2 \rangle^{\frac{1}{2}}_{exp}$ varies by a factor of three, the agreement between theory and experiment is rather good.

c(M)	1	0.5	0.1	0.04	0.01	0.005
$R_F/ S^2{}^{\frac{1}{2}}_{exp}$	1.395	1.448	1.323	1.329	1.321	1.370

Table II. The ratio $R_F/\langle S^2 \rangle^{\frac{1}{2}}_{exp}$ as a function of the ionic strength for KPSS

This implies that scaling theory[23] in which the asymptotic formula (4.9) is assumed, should be reasonable.

The intrinsic viscosity data can be explained as well. The theoretical formula we propose is:

$$M[\eta] = \alpha_\eta^3 \, \Phi\left(\frac{\ell}{2L_t}, \frac{d_H}{2L_t}\right) \ell^{3/2} (2L_t)^{3/2} \qquad (4.10)$$

The term $\ell^{3/2}\ (2L_t)^{3/2}$ is the hypothetical theta state expansion, α_η is a function of α_s (or $\alpha_{s,el}$) calculated by Weill and des Cloizeaux[52] and corrected by Francois et al.[54], and the ϕ function is a draining term for the wormlike chain model calculated by Yamakawa and Fujii[53]. The latter term depends on the reduced contour length $\ell/2L_t$ and the reduced hydrodynamic diamter $d_H\ /2L_t$. We assume d_H is constant and semi-empirical. From ref.[51] we derive that $d_H \simeq 1$ Å. As an example we have calculated $[\eta]$ at c=0.005M. The respective quantities are $\alpha_{s,el}$=1.607, α_D=1.313 (from ref.[54]), $\alpha_\eta^{\ 3}=\alpha_s^2\alpha_D$=3.389, $\ell/2L_t$=28.74, $d_H/2L_t$=0.0067, ϕ=1.084x10^{23} so that $[\eta]$=5.022 dℓ/g compared with the experiment $[\eta]$=4.41 dℓ/g which is certainly reasonably good agreement. We note that we have neglected the fact that actually non-asymptotic excluded-volume theory might have to be used but the corrections will only slightly lower the previous values. Finally, we point out that deriving persistence lengths from intrinsic viscosity data[55,56] is useless unless excluded-volume effects are also accounted for.

5. COLLAPSE OF WEAKLY CHARGED POLYELECTROLYTES

Relatively little work is expended in compressing flexible coils into sheets, tubes or spheres. Illuminating in this respect are some of the scaling theories reviewed by de Gennes[57]. If attractive forces between the segments or otherwise are externally induced, we may expect the chains to collapse dramatically. Polyelectrolyte coils are expecially suitable for studying such effects since one is able to alter accurately the strength of the attractive (or repulsive) interactions. This is exemplified by two recent calculations[58,59].

Khokhlov[58] has studied the collapse of very weakly charged polyions (i.e. A>>Q, e.g. A≃100 Å. In this case $A \gg L_p$ for a flexible polyelectrolyte and the concept of local stiffness is meaningless. On a local scale the chain is Gaussian because the widely spaced charges have virtually no influence on the configurations. Nevertheless, if we assume the solution to be sufficiently dilute so that Debye-Hückel screening can be neglected - there are very few counterions - the mean-square extension length must be proportional to the number of charges squared as argued by de Gennes et al.[60].

Thus, locally a so-called Gaussian blob of dimension B and containing g charges is built up. Successive blobs are strongly repelled by the Coulomb force thus forming a linear array of blobs. The free energy of repulsion between adjacent blobs must be of the order k_BT so that

$$g^2 \left(\frac{q^2}{D_B}\right) \simeq k_BT \tag{5.1}$$

The Gaussian blob has the obvious property

$$B^2 = g \left(\frac{A}{2L_p}\right) (2L_p)^2 \qquad (5.2)$$

The elongated conformation of the whole has a longitudinal size given by

$$\Lambda \simeq \frac{N}{g} \left(\frac{2L_p}{A}\right) B \qquad (5.3)$$

Eqs.(5.1) and (5.3) can be rewritten in terms of the known parameters

$$g \simeq \left(\frac{AL_p}{Q^2}\right)^{1/3} \qquad (5.4)$$

$$B \simeq \left(\frac{A^2 L_p^2}{Q}\right)^{1/3} \qquad (5.5)$$

$$\Lambda \simeq N \left(\frac{L_p^4\ Q}{A^2}\right)^{1/3} \qquad (5.6)$$

Eq.(5.6) shows that the coil is certainly not completely stretched and in fact has an elongation an order of magnitude smaller than the contour length.

In the above analysis only Coulomb repulsion was taken into account i.e. the system was considered to be at the theta temperature θ with respect to Van der Waals interactions. The latter can be switched on when $T<\theta$ with $\tau \equiv T-\theta/T<<1$. The Van der Waals forces exert a considerable influence on the conformations as has been reviewed before[61)].

Below a cririical temperature $T_r<\theta$, calculated by Khokhlov, the blobs collapse into much more compact globules. Again, the globules themselves are aligned more or less linearly because of the dominating Coulomb repulsion. Khokhlov derives the following expressions for the rodlike array of globules in the very dilute region. ($C_m << N\Lambda^{-3}$, C_m is the segment concentration).

$$g \simeq \frac{A\tau}{Q} \qquad (5.7)$$

$$B \simeq \left(\frac{v\ A^2}{L_p Q}\right)^{1/3} \qquad (5.8)$$

$$\Lambda \simeq N \left(\frac{v\ L_p^2 Q^2}{A^4}\right)^{1/3} \qquad (5.9)$$

Here v is defined by $-A_2/\tau$ where A_2 is the second virial coefficient pertaining to the segment-segment Van der Waals interaction.

A remarkable feature about eq.(5.9) is the rather strong dependence of Λ on A. The linear charge density increases as A increases. Eq.(5.6) does not show this anomaly which is caused only by the presence of the Van der Waals forces. One can therefore speculate whether counterion condensation occurs again just as in the case of highly charged polyelectrolytes. The last part of Khokhlov's paper is devoted to this problem and the answer is affirmative. Briefly, one can state that in the semi-dilute region ($g A L_p^{-1} B^{-3} \gg C_m \gg N\Lambda^{-3}$) and at a temperature lower than T_r (which in turn is lower than T_c) with $(\theta - T_r)/\theta \simeq (v/L_p AQ)^{1/3}$, counterions accumulate like an avalanche onto the rodlike globular polyions. It is a case of positive feedback: at T_r, the linear charge density is high enough for the counterions to start condensing. Hence, A increases, but according to eqs.(5.7)-(5.9), the linear charge density is also enlarged. This, in turn, implies yet more counterion accumulation, and so forth. Eventually, we end up with a dilute solution of fairly thick dense globular macromolecules. To date no experiments have been performed to test this interesting theory.

Another example of drastically collapsing chains within a different context, has been given by the author[59]. The cylindrically symmetrical electrostatic field of a stiff, highly charged rodlike polyanion like DNA attracts flexible polycations, a bound state appearing at a certain salt concentration $c_{s,1}$. When the salt concentration is decreased, the absorbed sheath collapses dramatically. If excluded-volume effects are neglected - a not unreasonable first-order approximation for weakly charged coils - an almost analytical calculation can be given. The derivation of the density profile of the adsorbed polycation turns out to be a turning point problem which can be solved by the Langer WKBJ approximation.

The decay length ζ which characterizes approximately the thickness of the bound layer is given by[59]

$$[C_1 A_K + a(1 - C_2 a^2 \kappa_1^2)^{\frac{1}{2}}] \; (\kappa_1 - \kappa) = C_3 A_K \zeta^{-1} \qquad (5.10)$$

where a is the radius of the cylindrical polyanion, C_1,C_2 and C_3 are constants, $\kappa_1^2 = 8\pi Q\, c_{s,1}$ and A_K is the Kuhn length of the flexible polycation. This equation is valid only for $\kappa_1 - \kappa$ defined within a narrow range. On the one hand the bound layer must be thick i.e. $\zeta \gg A$, on the other hand $\zeta \ll N_K^{\frac{1}{2}} A_K$. As can be seen, a small decrease in κ is sufficient to cause a considerable collapse of the sheath.

6. CONCLUDING REMARKS

An active field of research which we have not reviewed is scaling theory[23,60,62,63] which has already led to a number of interesting experimental developments[63-70]. However, the dynamics of polyelectrolytes without added salt is still unsolved for the very reason that no theory of entangled stiff chains exists. We have also disregarded dielectric and electro-optical properties but Mandel's review[7] partly alleviates this neglect. As a matter of fact his survey is an introduction to almost the whole field and provides good background for those interested in entering it. The reader should also take note of the very recent papers emphasizing the Poisson-Boltzman approach to polyelectrolyte solutions.[72-75]

In conclusion we can state that polyelectrolyte theory should still see further developments in the coming years. It is pertinent to realize that in 1976 several well-known people stated that not much new could be devised or measured. Fortunately, a number of workers have proven them wrong.

REFERENCES

1. M.Mandel in Polyelectrolytes, E.Sélegny, M.Mandel and U.P.Strauss ed., D.Reidel, Dordrecht, Holland (1974).
2. A.Katchalsky, Pure Appl.Chem., 26, 327 (1971).
3. F.Oosawa, Polyelectrolytes, Dekker, New York (1971).
4. A.H.MacGillivray and J.J.Winkleman Jr., J.Chem.Phys., 45, 2184 (1966).
5. G.S.Manning, J.Chem.Phys., 51, 924 (1969).
6. K.Iwasa, D.A.McQuarrie and J.C.T.Kwak, J.Phys.Chem.82, 1979 (1978).
7. D.Stigter, J.Phys.Chem., 82, 1417 (1978).
8. M.Fixman, J.Chem.Phys., 70, 4995 (1979).
9. M.A.Lampert and R.S.Crandall, Chem.Phys.Lett., 72, 481 (1980).
10. H.Maeda, N.Saitô and W.H.Stockmayer, Polym.J.2, 94 (1971).
11. L.D.Landau and E.M.Lifshitz, Statistical Physics Part I, 3rd. edition, § 127, Pergamon, Oxford (1980).
12. H.Yamakawa, Modern Theory of Polymer Solutions, Harper and Row, New York (1971).
13. H.Yamakawa, Pure Appl.Chem., 46, 135 (1976).
14. H.Yamakawa and M.Fujii, J.Chem.Phys., 64, 5222 (1976).
15. H.Yamakawa, Macromolecules, 10, 692 (1977).
16. A.Miyake and Y.Hoshino, J.Phys.Soc.Jap., 46, 1324 (1979).
17. J.M.Bailey, Macromolecules, 12, 91 (1979).
18. T.Odijk, J.Polym.Sci.,Polym.Phys.Ed., 15, 477 (1977).
19. J.Skolnick and M.Fixman, Macromolecules, 10, 944 (1977).
20. T.Odijk, Polymer, 19, 989 (1978).
21. T.Odijk and A.C.Houwaart, J.Polym.Sci., Polym.Phys.Ed., 16, 627 (1978)
22. M.Fixman and J.Skolnick, Macromolecules, 11, 863 (1978).
23. T.Odijk, Macromolecules, 12, 688 (1979).
24. T.Odijk and M.Mandel, Physica, 93A, 298 (1978).
25. D.M. Soumpasis, J.Chem.Phys., 69, 3190 (1978).
26. J.M.Schurr and S.A.Allison, Biopolymers, 20, 251 (1981).
27. G.S.Manning, Quart.Rev.Biophys., 11, 179 (1978).
28. K.F.Freed, J.Chem.Phys., 54, 1453 (1971).
29. M.le Bret, C.R.Acad.Sci., Paris, II, 292, 291 (1981).
30. J.A.Schellman, Biophys.Chem., 11, 321 (1980).
31. R.E.Harrington, Biopolymers, 17, 919 (1978).
32. V.Rizzo and J.A.Schelleman, Biopolymers, 20, 2143 (1981).
33. P.J.Hagerman, Biopolymers, 20, 1503 (1981).
34. M.Mandel and J.Schouten, Macromolecules, 13, 1247 (1980).
35. N.Borochov, H.Eisenberg and Z.Kam, Biopolymers, 20, 231 (1981).
36. Z.Kam, N.Borochov and H.Eisenberg, Biopolymers, 20, 2671 (1981).
37. G.S.Manning, Biopolymers, 18, 3111 (1981).
38. G.Weill in:Molecular Electro-Optics, S.Krause ed., Plenum, New York (1981).
39. G.Maret and G.Weill, private communication.
40. D.P.Millar, R.J.Robbins and A.H.Zerwail, J.Chem.Phys., 76, 2080 (1982).
41. B.H.Zimm, J.Chem.Phys., 14, 164 (1946).
42. L.Onsager, Ann.N.Y.Acad.Sci., 51, 627 (1949).
43. H.Yamakawa and W.H.Stockmayer, J.Chem.Phys., 57, 2843 (1972).
44. Y.Oono, J.Phys.Soc.Jap., 39, 25 (1975).
45. S.F.Edwards, J.Phys.A., 8, 1171 (1975).
46. S.F.Edwards, Proc.Phys.Soc.Lond., 85, 613 (1965).
47. P.G. de Gennes, Phys.Lett.A,38, 399 (1972).
48. M.K.Kosmas and K.F.Freed, J.Chem.Phys., 68, 4878 (1978).

49. T.Odijk, J.Phys.A., 11, L153 (1978).
50. D.M.Soumpasis and K.H.Bennemann, Macromolecules, 14, 50 (1981).
51. A.Raziel and H.Eisenberg, Isr.J.Chem., 11, 183 (1973).
52. G.Weill and J. des Cloizeaux, J.Phys.(France), 40, 99 (1979).
53. H.Yamakawa and M.Fujii, Macromolecules,7, 128 (1974).
54. J.François, T.Schwarz and G.Weill, Macromolecules, 13, 564 (1980).
55. S.Miyamoto, Makromol.Chem., 182, 559 (1981).
56. T.Shimuzu, Polymer, 22, 231 (1981).
57. P.G. de Gennes, Scaling Concepts in Polymer Physics, Cornell University Press, Ithaca, New York (1979).
58. A.R.Khokhlov, J.Phys.A13, 979 (1980).
59. T.Odijk, Macromolecules, 13, 1542 (1980).
60. P.G. de Gennes, P.Pincus, R.M.Velasco and F.Brochard, J.Phys.(France), 37, 1461 (1976).
61. I.M. Lifshitz, A.Yu Grosberg and A.R.Khokhlov, Rev.Mod.Phys., 50, 683 (1978).
62. P.Pfeuty, J.Phys.Coll. C2.(France), 39, 149 (1978).
63. J.Hayter, G.Janninck, F.Brochard-Wyart and P.G. de Gennes, J.Phys.Lett.France, 41, L451 (1980).
64. C.Williams et al., J.Poly.Sci.,Poly.Lett.Ed., 17, 379 (1970).
65. M.Nierlich et al., J.Phys.France, 40, 701 (1979).
66. R.S.Koene, H.W.J.Smit and M.Mandel, Chem.Phys.Lett., 74, 176 (1980).
67. J.Roots and B.Nyström, Polymer, 22, 573 (1981).
68. F.Grüner, W.P.Lehmann, H.Fahlbusch and R.Weber, J.Phys.A 14, L307 (1981).
69. M.Levij, J. de Bleijser and J.C.Leyte, Chem.Phys.Lett., 83, 183 (1981).
70. M.Levij, J. de Bleijser and J.C.Leyte, Chem.Phys.Lett., 87, 34 (1982).
71. M.Mandel in:Molecular Electro-Optics, S.Krause ed., Plenum, New York (1981).
72. D.Stigter, Macromolecules, 15, 635 (1981).
73. M.le Bret, J.Chem.Phys., 76, 6243 (1982).
74. M.Fixman, J.Chem.Phys., 76, 6346 (1982).
75. W.B.Russel, J.Polym.Sci.,Polym.Phys.Ed., 20, 1233 (1982).

DYNAMICAL PROPERTIES OF CHARGED SPHERICAL BROWNIAN PARTICLES

R. Klein and W. Hess
Fakultät für Physik
Universität Konstanz
D-7750 Konstanz, Germany

I. Introduction

Suspensions of large molecules in solvents consisting of small molecules are of interest in various fields. Whereas polymer physics considers mainly flexible macromolecules, rigid macroparticles are relevant for colloid science and are of importance in biophysical systems. Their static and dynamical properties can be investigated by light scattering techniques; in particular, light beating spectroscopy is a valuable tool to determine the slow dynamics of the macroparticles[1]. Whereas in very dilute solutions every macromolecule is a free Brownian particle, the situation becomes more involved at concentrations such that the individual particles start to interact. The main object of this paper is the investigation of systems of interacting Brownian particles. Experimental results exist for many systems of this kind[2]; a few examples are proteins and viruses in solution, biological or artificial micelles and microemulsions. Depending on the type of particles and on the solvent, different kinds of interactions among the macroparticles are of primary importance. Here, we will restrict ourselves to charged spherical macroparticles in solution. Various proteins and viruses in aqueous solution are of this type. For these solutions there is a model system which has been investigated in great detail: polystyrene spheres in water. This system can be prepared in a nearly monodisperse way, the spheres scatter light much more strongly than the solvent, so that the measured light scattering spectrum arises only from the interacting Brownian particles.

After a short review of the static properties of such systems the main part of this paper is devoted to the development of a theory for the dynamical properties of charged interacting spherical Brownian particles. Starting from a Fokker-Planck equation as the basic transport equation the generalized hydrodynamics of the system is developed in analogy to the theory of simple liquids. Basic differences arise because of the large friction of the macroions in the solvent.

The dynamical structure factor is calculated by a method which is equivalent to a two-step memory function procedure, the first memory function being a generalized diffusion function and the second one the longitudinal dynamic viscosity of the liquid of Brownian particles. It is shown how this function can be obtained from the experimental light scattering spectrum. It is also calculated by a mode-mode coupling method and satisfactory agreement is found between theory and experiment. In a final part we present results for single particle properties such as the self-diffusion coefficient and the velocity autocorrelation function.

II. Static Structure Factor

Static light scattering experiments on polystyrene spheres in aqueous solution show clearly that these systems are strongly interacting even at rather low concentrations[3,4,5]. The systems investigated in ref. 5 cover a concentration range of volume fractions between $\phi = 0.0015$ and $\phi = 0.0075$, where the mean interparticle distance is larger than 10 times the particle diameter. The experiments show liquid-like structure factors $S(\underline{k})$ with well-developed first peaks, where $S(k_{max}) \approx 1.5$ for the lowest concentration and $S(k_{max}) \approx 2.8$ for the highest one. Here k_{max} is the wave number at which $S(k)$ has its first maximum. The extrapolations of $S(k)$ for $k \to 0$ show that $S(0)$ is decreasing with increasing concentration.

Models for $S(k)$ have been developed, which describe the solution of macroparticles as a one-component plasma of ions of finite size. The counter-ions are treated as a uniform neutralizing background, which determines the screening of the ion-ion interactions, which is taken to be a repulsive screened Coulomb interaction. The attractive van der Waals interactions are assumed to be negligible since the Coulomb interactions are in most cases larger than k_BT. The two-particle potential is taken as[6]

$$U(r_{ij}) = \begin{cases} \pi \, \varepsilon_0 \, \varepsilon \, d^2 \, \psi_0^2 \, \frac{1}{r_{ij}} \exp[-\kappa(r_{ij} - d)] \; ; & r_{ij} > d \\ \infty \; ; \quad r_{ij} < d & \end{cases} \tag{1}$$

where ε is the dielectric constant of the solvent, d the diameter of the macroions, ψ_0 the surface potential, κ the Debye-Hückel screening parameter, and $r_{ij} = |\underline{r}_i - \underline{r}_j|$. Hayter and Penfold[7] have calculated $S(k)$ on the basis of the pair-potential (1) using the mean-spherical approximation (MSA). This procedure works well for rather concentrated systems like concentrated micellar solutions, but it fails at low densities, such as the highly charged polystyrene systems. This difficulty has been overcome subsequently by Hansen and Hayter[8] using a rescaling procedure. These authors were able to reproduce the experimental structure factors obtained by Brown et al.[3] quite well. Only in the long-wavelength limit the calculated $S(k)$ is systematically lower than the measured values. But it should also be mentioned that experimental uncertainties exist particularly at small values of k due to multiple scattering.

We have used the method of Hansen and Hayter to obtain $S(k)$ for the polystyrene solutions investigated by Grüner and Lehmann[5]. The value of the surface potential ψ_0, which is difficult to determine experimentally, is obtained by fitting the height of the main peak of $S(k)$ for one concentration. Keeping this value fixed for all other samples, good agreement was obtained with the experimentally determined structure factors; again $S(k \to 0)$ is found to be smaller than the extrapolations of the experimental curves suggest. Since $S(k)$ will lateron be used as input in the calculations of dynamical properties and since it is well-known that the values of S(k) for small k are of particular importance, we have corrected the long-wavelength part of S(k) by a phenomenological method first proposed by Minoo et al.[9] in the context of the electrical properties of liquid metals. The structure factor is written as

$$S(k) = \gamma_1(k)\, S(0) + \left[1 - \gamma_2(k)\right] S_{HH}(k) \quad , \tag{2}$$

where $S_{HH}(k)$ is the Hansen-Hayter structure factor. The functions $\gamma_{12}(k)$ are strongly decaying functions of k, and S(0) is chosen in agreement with the extrapolated experimental values. An independent determination of S(0) through the measurement of the osmotic compressibility would be very helpful in clearing up the uncertainties which are introduced by this phenomenological method. The method just outlined for obtaining S(k) was used to calculate static structure factors for some of the systems investigated in Ref. 5, the results are shown in Fig. 1. The systems are characterized by the dimensionless coupling constant $\gamma = \beta\, \pi\, \varepsilon_o\, \varepsilon\, d\, \psi_o^2 \exp(d\kappa) = 250$.

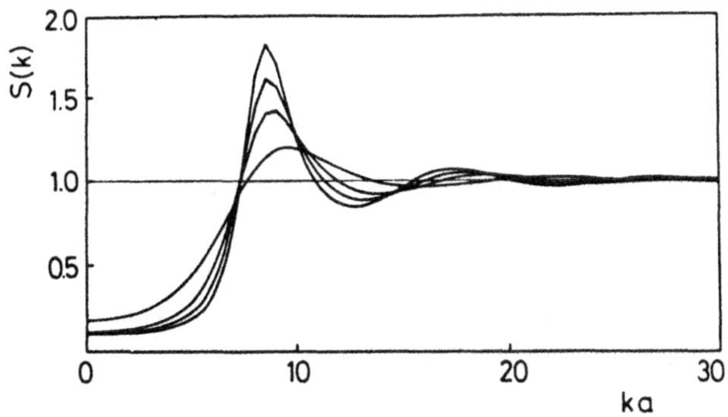

Fig. 1. Static structure factors for polystyrene spheres[5] of diameter 900 Å, ψ_o = 73 mV, κ^{-1} = 5000 Å and volume concentration $\phi \cdot 10^3$ = 0.1;0.5;1.5;4.5. The interparticle spacing is $a = d/(2\phi^{1/3})$.

III. Dynamic Structure Factor

As the static structure factor very clearly shows the dilute aqueous solutions of highly charged polystyrene spheres are strongly interacting systems. Because of the large difference of the masses of solvent and solute particles the time scales for the dynamics are very different. Restricting ourselves to the time scale which is experimentally accessable by light beating spectroscopy we model the solutions as systems of interacting Brownian particles. The momenta $\underline{p}_i$ and coordinates $\underline{r}_i$ of the macroions are the dynamical variables and the solvent is described as a continuous medium which produces friction and mediates a hydrodynamic interaction between different macroparticles. The basic equation is then the Fokker-Planck equation[10,11] for the distribution function in the phase space of the macroions, $f(\Gamma,t) \equiv f(\underline{p}_1,\ldots,\underline{p}_N;\underline{r}_1,\ldots,\underline{r}_N;t)$

$$\frac{\partial}{\partial t} f(\Gamma,t) = \hat{\Omega}(\Gamma)\, f(\Gamma,t) \qquad (3)$$

where $\hat{\Omega}$ is the Fokker-Planck operator

$$\hat{\Omega}(\Gamma) = -\sum_i \left(\frac{1}{m}\,\underline{p}_i\cdot\frac{\partial}{\partial \underline{r}_i} + \underline{F}_i\cdot\frac{\partial}{\partial \underline{p}_i}\right) + \sum_{i,j} \frac{\partial}{\partial \underline{p}_i}\cdot\underline{\underline{\zeta}}_{ij}\cdot\left(k_B T\,\frac{\partial}{\partial \underline{p}_j} + \frac{1}{m}\,\underline{p}_j\right). \qquad (4)$$

$\underline{F}_i$ denotes the direct forces exerted on particle i by all other particles and the $\underline{\underline{\zeta}}_{ij}$ are cross friction tensors describing the hydrodynamic interactions. For the special system of highly charged particles in aqueous solution at rather low concentration the hydrodynamic interaction can be neglected since the strong Coulomb repulsion keeps the particles far apart. The situation is quite different if the solvent is less polar like benzene; then the particles do not become polyelectrolytes and the hydrodynamic interactions are very important[12]. Without hydrodynamic interaction

$$\underline{\underline{\zeta}}_{ij} = \zeta_o\, \delta_{ij}\, \mathit{1} \qquad (5)$$

where $\mathit{1}$ is the unit tensor and ζ_o denotes the friction coefficient of one particle. The most important phase space variables are the Fourier components of the density fluctuations

$$\hat{c}(\underline{k}) = \sum_i \exp(-i\,\underline{k}\cdot\underline{r}_i) - \left\langle \sum_i \exp(-i\,\underline{k}\cdot\underline{r}_i)\right\rangle_o , \qquad (6)$$

where the index zero indicates an equilibrium average, and the current fluctuations

$$\hat{\underline{j}}(\underline{k}) = \sum_i \frac{1}{m}\,\underline{p}_i \exp(-i\,\underline{k}\cdot\underline{r}_i) . \qquad (7)$$

Their time derivatives are given by $\hat{\Omega}^+\,\hat{c}(\underline{k})$ and $\hat{\Omega}^+\,\hat{\underline{j}}(\underline{k})$, where $\hat{\Omega}^+$ is the Hermitean adjoint Fokker-Planck operator. Using (4), (6) and (7), one finds that only for the concentration fluctuations the time derivative goes to zero in the long wavelength limit. This expresses the fact that only the total number of macroparticles is a conserved quantity whereas momentum and energy of the subsystem of interacting Brownian particles are not conserved because of friction in the solvent.

The conserved quantities are now used to split any dynamical variable $\hat{A}$ into a part, which is in local equilibrium and a rest $\hat{A}'$:

$$\hat{A}(\underline{k}) = \frac{\langle \hat{A}(\underline{k})\, \hat{c}(-\underline{k})\rangle_o}{\langle \hat{c}(\underline{k})\, \hat{c}(-\underline{k})\rangle_o}\, \hat{c}(\underline{k}) + \hat{A}'(\underline{k}). \qquad (8)$$

Applying this decomposition to the time derivative of the current fluctuations, the local equilibrium part is identified with the pressure fluctuations $\hat{p}(k)$ and the rest is a fluctuating force density $\hat{\underline{f}}(\underline{k})$:

$$m\,\hat{\Omega}^+\, \underline{j}(\underline{k}) = -\, i\, \underline{k}\, \hat{p}(\underline{k}) + \hat{\underline{f}}(\underline{k}) \qquad (9)$$

with

$$\hat{p}(\underline{k}) = \frac{k_B T}{S(\underline{k})}\, \hat{c}(\underline{k}) \; ; \qquad (10a)$$

$$S(\underline{k}) = \langle \hat{c}(\underline{k})\, \hat{c}(-\underline{k})\rangle_o \qquad (10b)$$

$$\hat{\underline{f}}(\underline{k}) = i\ \underline{k}\cdot\hat{\underline{\underline{\sigma}}}(\underline{k}) - \zeta_0\ \hat{\underline{j}}(\underline{k}) \tag{10c}$$

$\hat{\underline{\underline{\sigma}}}(\underline{k})$ is the (viscous) stress tensor of the "liquid of interacting Brownian particles" and formally identical to the expression obtained in the theory of simple liquids. The second term in (10c) is a friction force density $\hat{\underline{f}}^F(\underline{k})$, which is finite in the long wavelength limit.

The equations of motion of the macrovariables corresponding to (6) and (7) are the continuity equation

$$\frac{\partial c(\underline{k},t)}{\partial t} = -\ i\ \underline{k}\cdot\underline{j}(\underline{k},t) \tag{11}$$

and a Navier-Stokes equation with an external force density, which represents the friction due to the solvent

$$m\frac{\partial \underline{j}(k,t)}{\partial t} = -\ i\ \underline{k}\ p(\underline{k},t) + i\ \underline{k}\cdot\underline{\underline{\sigma}}(\underline{k},t) + \underline{f}^F(\underline{k},t)\,. \tag{12}$$

In order to close this set of equations there are two traditional procedures. The first one looks at the problem as a diffusion problem; accordingly, the current is written as arising from gradients in the particle distribution

$$\underline{j}(\underline{k},t) = -\int_0^t dt'\ \underline{\underline{D}}(\underline{k},t-t')\cdot\ \underline{\nabla}\ c(\underline{k},t'). \tag{13}$$

The second one is common in hydrodynamics: One assumes that the total non-equilibrium force fluctuations $\underline{f}(\underline{k},t)$ are produced by current fluctuations

$$\underline{f}(\underline{k},t) = -\int_0^t dt'\ \underline{\underline{\zeta}}(\underline{k},t-t')\cdot\underline{j}(\underline{k},t'). \tag{14}$$

These relations introduce generalized transport coefficients, a diffusion tensor $\underline{\underline{D}}(\underline{k},t)$ and a friction tensor $\underline{\underline{\zeta}}(\underline{k},t)$.

The dynamic structure factor

$$S(\underline{k},t) = \langle c(\underline{k},t)\ c(-\underline{k},0)\rangle_0 \tag{15}$$

can now be expressed by the transport coefficient which were just introduced. From (13) and (11) the Laplace transform of (15) becomes

$$\tilde{S}(\underline{k},z) = \frac{S(\underline{k})}{z + \tilde{D}(\underline{k},z)k^2} \quad ; \quad \tilde{D}(\underline{k},z)k^2 = \underline{k}\cdot\tilde{\underline{\underline{D}}}(\underline{k},z)\cdot\underline{k}\ . \tag{16}$$

On the other hand, from (12), (14) and (10a) the current autocorrelation function can be calculated, which is through (11) simply related to $S(\underline{k},t)$. As a result,

$$\tilde{S}(\underline{k},z) = \frac{S(\underline{k})}{z + \dfrac{k^2 c_T^{\,2}(\underline{k})}{z + \frac{1}{m}\tilde{\zeta}_{\shortparallel}(\underline{k},z)}} \tag{17}$$

where $c_T^{\,2}(k) = k_B T/\big(m\ S(k)\big)$ is the squared isothermal sound velocity of the system of macroions.

Comparing (16) and (17) gives the generalized Stokes-Einstein relation

$$\tilde{D}(\underline{k},z) = \frac{c_T^{\,2}(k)}{z + \frac{1}{m}\tilde{\zeta}_{\parallel}(\underline{k},z)} \tag{18}$$

Another way to look at the closure relations (13) and (14) and the transport coefficients which they introduce is to observe from (16) that $\tilde{D}(\underline{k},z)$ is essentially the memory function of the concentration autocorrelation function and that (18) shows that $\tilde{\zeta}_{\parallel}(\underline{k},z)$ is closely related to the memory function of $\tilde{D}(\underline{k},z)$. These results can also be obtained by applying a projection operator formalism directly to the Fokker-Planck equation.

Whereas this latter approach yields microscopic expressions for the transport coefficients, the phenomenological introduction through eqs. (13) and (14) calls for a separate calculation of one of these coefficients; the other one can then be obtained from eq. (18). We have calculated the generalized friction tensor $\zeta_{\alpha\beta}(\underline{k},t)$ by using linear response theory. Introducing an external force field $\underline{E}$ gives rise to an additional term in the Fokker-Planck operator. To linear order in $\underline{E}$ the response of the dynamical variable $\hat{A}(\underline{k})$ is given by

$$\begin{aligned} \langle A(\underline{k},t)\rangle &= \sum_{\alpha=1}^{3}\int_{o}^{t} dt' \,\frac{\beta}{V} \langle \hat{A}(\underline{k}) \exp|\,\hat{\Omega}(t-t')\,|\,\hat{j}_{\alpha}(-\underline{k})\rangle_{o} \cdot E_{\alpha}(\underline{k},t') \\ &\equiv \sum_{\alpha=1}^{3}\int_{o}^{t} dt' \,\psi_{A,E_{\alpha}}(\underline{k},t-t')\, E_{\alpha}(\underline{k},t') \,, \end{aligned} \tag{19}$$

where the last equality defines a generalized susceptibility $\psi_{A,E_{\alpha}}(\underline{k},t)$. In this way it can be shown[13] that the friction tensor can be expressed as a ratio of two such susceptibilities

$$\tilde{\zeta}_{\alpha\beta}(\underline{k},z) = -\frac{\tilde{\psi}_{f_{\alpha},E_{\beta}}(\underline{k},z)}{\tilde{\psi}_{j_{\beta},E_{\beta}}(\underline{k},z)} \tag{20}$$

Using the explicit form of the Fokker-Planck operator, eq. (2), the following result is obtained for the longitudinal friction function

$$\tilde{\zeta}_{\parallel}(\underline{k},z) = \zeta_{o} + (k^2/c)\,\tilde{\eta}_{\parallel}(\underline{k},z) \,, \tag{21}$$

where $\eta_{\parallel}$ denotes the longitudinal viscosity of the liquid consisting of the interacting Brownian particles[14],

$$\tilde{\eta}_{\parallel}(\underline{k},z) = \langle\hat{\sigma}_{zz}(\underline{k})\,(z - \hat{Q}\,\hat{\Omega}\,\hat{Q})^{-1}\,\hat{\sigma}_{zz}(-\underline{k})\rangle_{o} \tag{22}$$

Here, $\hat{Q}$ is a projection operator

$$\hat{Q} = 1 - \hat{P}_{c} - \hat{P}_{j} \,, \tag{23}$$

where $\hat{P}_c$ projects onto $\hat{c}(\underline{k})$ and $\hat{P}_j$ projects onto $\hat{\underline{j}}(\underline{k})$.

IV. Limiting behavior of $S(\underline{k},t)$ and Viscoelastic Approximation

The short-time behavior of $S(\underline{k},t)$ can be characterized by the first few frequency moments $m_n(\underline{k})$ defined by

$$S(\underline{k},t) = \sum_{n=0}^{\infty} \frac{1}{n!} m_n(k)\, t^n . \tag{24}$$

From (17) and (21) one obtains

$$\begin{aligned} &m_1(\underline{k}) = 0 \;;\; m_2(\underline{k}) = -k^2/(m\beta) \;;\; m_3(\underline{k}) = k^2\, \zeta_o/(m^2\beta) \\ &m_4(\underline{k}) = \frac{k^2}{m\beta} \{k^2 (c_T^{\,2}(k) + \frac{1}{cm}\, \eta_{\shortparallel}(\underline{k},0)) - (\frac{\zeta_o}{m})^2\} . \end{aligned} \tag{25}$$

The initial value of the longitudinal viscosity $\eta_{\shortparallel}(\underline{k},0)$ has the same form as in simple liquids

$$\eta_{\shortparallel}(\underline{k},0) = c\, \{3\, k_B T + \frac{1}{k^2 N} \sum_{i,j} < \frac{\partial^2 U_N}{\partial z_i \partial z_j}\, e^{-ik(z_i - z_j)} >_o - \frac{k_B T}{S(k)} \} ,$$

where U_N is the total interaction potential.

The continued fraction representation, which reproduces correctly the first four moments, is

$$\tilde{S}(\underline{k},z) = \cfrac{S(k)}{z + \cfrac{c_T^2(\underline{k})\, k^2}{z + \frac{\zeta_o}{m} + \cfrac{k^2\, \eta_{\shortparallel}(\underline{k},0)/cm}{z + \dots}}} . \tag{26}$$

Comparison with (17) shows that this approximation corresponds to replacing the exact function $\tilde{\zeta}_{\shortparallel}(\underline{k},z)$ by $\zeta_o + k^2\, \eta_{\shortparallel}(\underline{k},0)/(z\, c\, m)$. If one uses this approximation which is appropriate at short times, in the equation for the longitudinal current $j_{\shortparallel}(\underline{k},t)$, one obtains

$$m \frac{\partial^2 j_{\shortparallel}(\underline{k},t)}{\partial t^2} + \zeta_o \frac{\partial j_{\shortparallel}(\underline{k},t)}{\partial t} + \frac{k^2}{c} \left[\tfrac{4}{3}\, G^{\infty}(k) + K^{\infty}(k)\right] j_{\shortparallel}(\underline{k},t) = 0 . \tag{27}$$

This shows that the systems behaves, at short times, like a damped harmonic system with high frequency elastic constants

$$\tfrac{4}{3}\, G^{\infty}(k) + K^{\infty}(k) = m\, c\;\; c_T^{\,2}(k) + \eta_{\shortparallel}(\underline{k},0) . \tag{28}$$

The opposite limit is the hydrodynamic one. According to the earlier discussion concerning the number of conserved quantities, only one hydrodynamic pole is expected which is the mass diffusion pole. From (16), (18) and (21) the dynamic structure factor becomes a simple exponential, the exponent being $D_c k^2$, where

$$D_c = \tilde{D}(0,0) = D_o/S(0) \tag{29}$$

is the collective or mass diffusion coefficient; $D_o = k_B T/\zeta_o$ is the diffusion constant of a free macroion. The result (29) has earlier been obtained[15] by describing the system by a Smoluchowski equation. Since S(0) is about 0.1 for the particular systems considered in Sect. II, $D_c \approx 10\ D_o$.

Whereas the friction function of a simple liquid vanishes as k^2 in the hydrodynamic limit, eq. (21) shows that it is finite for a Brownian system. The two propagating sound modes of the simple liquid therefore become a central line; in the hydrodynamic limit our system is an overdamped liquid. The light scattering experiments show indeed that $S(\underline{k},t)$ is a simple exponential in time for small wavevectors in agreement with (29).

One way to approximately interpolate between the two limiting cases is the viscoelastic approximation. Writing

$$\tilde{\eta}_{\parallel}(\underline{k},z) = \frac{\eta_{\parallel}(\underline{k},0)}{z + \tau^{-1}(\underline{k})} \quad ; \quad \tau(\underline{k}) = \tilde{\eta}_{\parallel}(\underline{k},0)/\eta_{\parallel}(\underline{k},0) \tag{30}$$

reproduces both limits.

V. Strongly overdamped system

So far the dynamics of the system of interacting macroions has been quite general on the time scale of the Fokker-Planck equation, which is applicable as long as the solvent dynamics has not to be taken into account. Since the shortest correlation times of a dynamical light scattering experiment are of the order of 10^{-5} to 10^{-6} sec, important simplifications of our general results can be introduced. As an introduction to this procedure let us consider a simplified example and neglect for the moment the viscosity term in (21) as compared to ζ_o. In this case, $S(\underline{k},z)$ has two poles at

$$z_{1,2} = -\zeta_o/(2m) \pm \left[(\zeta_o/2m)^2 - c_T^{\,2}(k)\, k^2\right]^{1/2}$$

As long as $\zeta_o/m \ll c_T k$, there are two weakly damped sound waves. In the opposite limit, however, there are two widely separated poles at $z_1 = -\zeta_o/m \equiv -\tau_B^{-1}$ and $z_2 = -\left(c_T(k)k\right)^2\tau_B \ll z_1$. Moreover, the residue of $S(\underline{k},z=z_1)$ is very small compared to the residue at z_2. The pole at z_1 describes fast momentum relaxation due to the frequent encounters with the solvent, whereas the pole at z_2 is due to slow spatial diffusion. Since τ_B for the polystyrene spheres is of the order of 10^{-8}sec the light scattering experiments are always in the overdamped regime. Using now the complete expression (21) in (17) we can neglect z compared to τ_B^{-1} and write

$$\tilde{S}(\underline{k},z) = \frac{S(\underline{k})}{z + \dfrac{m\, c_T^{\,2}(k)\, k^2}{\zeta_o + (k^2/c)\, \tilde{\eta}_{\parallel}(\underline{k},z)}} \tag{31}$$

It should be kept in mind that this expression only applies at times longer than τ_B. On this new time scale, which is the time scale of light scattering experiments, the short-time behavior is, of course, quite different than before. One way to analyse the experimental results is to plot $\log|S(\underline{k},t)/S(\underline{k})|$ versus t, from which the first few cumulants can in principle be obtained. The cumulant expansion is

$$S(\underline{k},t) = S(\underline{k}) \exp\left[- \sum_{n=1}^{\infty} \frac{1}{n!} \mu_n(k)\, t^n\right] . \tag{32}$$

Using (31) one obtains

$$\begin{aligned} \mu_1(k) &= D^{\infty}(k)\, k^2 \quad ; \quad D^{\infty}(\underline{k}) = D_o/S(k) \\ \mu_2(k) &= - \mu_1(k)\, k^2\, \eta_{\parallel}(\underline{k},0)/(c\, \zeta_o) . \end{aligned} \tag{33}$$

Therefore, the first cumulant provides an experimental determination of the static structure factor $S(\underline{k})$ by a dynamical experiment. Furthermore, since $\mu_2(k) < 0$, there is always an upward curvature of $\log|S(\underline{k},t)/S(\underline{k})|$ versus t. Both results agree with the experimental results[2-5], the latter property arising from the viscoelastic behavior of the system of Brownian particles. Clearly, $\mu_2(k \to 0) = 0$. The fact that $\mu_1(k)$, eq. (33), is in agreement with experimental results for polystyrene spheres in aqueous solution proves that hydrodynamic interactions are negligible for these systems. Instead of analyzing the experiments by determining the second cumulant $\mu_2(k)$, it seems advantageous to introduce a mean relaxation time of $S(\underline{k},t)$

$$\begin{aligned} \tau(\underline{k}) &= \frac{1}{S(\underline{k})} \int_0^{\infty} dt\, S(\underline{k},t) \\ &= \frac{\zeta_o + k^2\, \tilde{\eta}_{\parallel}(\underline{k},0)/c}{k_B T\, k^2/S(\underline{k})} \end{aligned} \tag{34}$$

If $S(\underline{k},t)$ were a simple exponential function for all k , we would have $\tau(k)^{-1} = \mu_1(k)$. We introduce

$$\Delta(k) = \frac{\mu_1(k) - \tau(k)^{-1}}{\mu_1(k)} \tag{35}$$

as a measure for the deviation of $S(\underline{k},t)$ from an exponential. Using (33) and (34)

$$\Delta(k) = \frac{k^2\, \tilde{\eta}_{\parallel}(\underline{k},0)/(c\, \zeta_o)}{1 + k^2\, \tilde{\eta}_{\parallel}(\underline{k},0)/(c\zeta_o)} \tag{36}$$

The experimental determination of $\Delta(k)$ therefore gives the time integral of the dynamic longitudinal viscosity

$$\tilde{\eta}_{\parallel}(\underline{k},0) = \int_0^{\infty} dt\, \eta_{\parallel}(\underline{k},t). \tag{37}$$

This quantity is calculated by mode-mode coupling approximation. Starting from (22), the projection operator $\hat{Q}$ is replaced by bilinear products of concentration fluctuations and the resulting four-point correlation functions are approximated by the sum of bilinear products of two-point correlations. The result is

$$\eta_{\parallel}(\underline{k},t) = \frac{2\beta c^2}{(2\pi)^3} \int d^3k'\, \frac{1}{k^2}\, g_z(\underline{k},\underline{k}')\, g_z(-\underline{k},-\underline{k}')\, S(\underline{k}_1,t)\, S(\underline{k}_2,t) \tag{38}$$

where

$$g_z(\underline{k},\underline{k}') = \frac{-i\ k_BT}{2\ S(\underline{k}_1')\ S(\underline{k}_2')} \left[k_{1z}'\ h(\underline{k}_1') + k_{2z}'\ h(\underline{k}_2')\right] \tag{39}$$

$$\underline{k}_{1,2} = \frac{1}{2}\,\underline{k} \mp \underline{k}' \quad ; \quad c\ h(\underline{k}) = 1 - S(\underline{k}) \tag{40}$$

Eqs. (31) and (38) are a closed system of equations to determine the dynamic structure factor and the longitudinal viscosity function. Instead of solving this system self-consistently we have approximated $S(\underline{k},t)$ in (38) by the mean-field expression $S^{MF}(\underline{k},t) = S(k)\cdot\exp\{D^{\infty}(k)k^2t\}$. A particularly sensitive comparison between theory and experiment is provided by $\Delta(\underline{k})$, the deviation of $S(\underline{k},t)$ from a single exponential function in time.

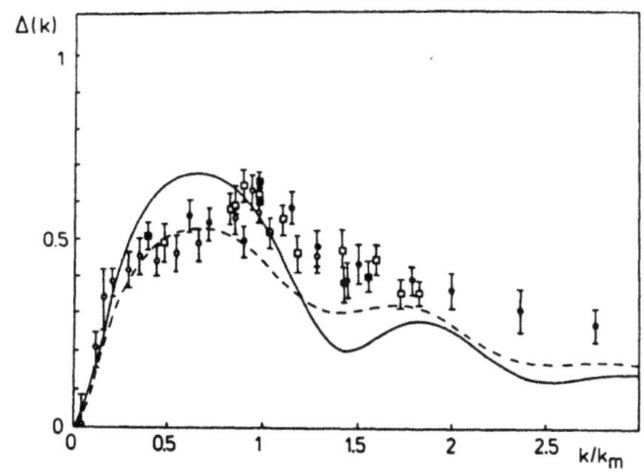

Fig. 2. Comparison of theoretical results, eq. (36), with experiments[5]. Full line is obtained with S(k) as calculated in Sect. II, broken line corresponds to the use fo the experimentally determined S(k).

Fig. 2 shows a comparison of our theoretical results with the experimentally determined[5] $\Delta(k)$. It seems somewhat surprising that the experiments yield values for $\Delta(k/k_{max})$ which are more or less independent of concentration. Using the concentration dependent S(k) as obtained in Sect. II, we have calculated $\Delta(k)$ as a function of concentration. It is found that after a rapid increase at very small volume fractions, $\Delta(k)$ is only slightly increasing in the concentration range investigated experimentally in Ref. 5, thereby explaining the apparent concentration independence of the data.

VI. Self-diffusion

The calculation of the self-diffusion coefficient D_s and of related quantities can be performed much in parallel to what has been done in the preceeding part of this paper. The correlation function of central importance is

$$G(\underline{k},t) = \left\langle \exp\left[-i\ \underline{k}\cdot\{\underline{r}_1(t) - \underline{r}_1(0)\}\right]\right\rangle_0 \tag{41}$$

where $\underline{r}_1(t)$ is the coordinate of the tagged particle. From $G(\underline{k},t)$ one can calculate the mean-square displacement

$$W(t) = -\frac{1}{2}\frac{\partial^2}{\partial k^2}\ G(k,t)\Big|_{k=0} \tag{42}$$

and the velocity auto-correlation function

$$Z(t) = - \lim_{k\to 0} \frac{\partial^2}{\partial t^2} \frac{1}{k^2} G(k,t) \,. \tag{43}$$

One can show[13] that the Laplace transform of (41) can be written as

$$\tilde{G}(k,z) = \left[z + \tilde{D}_s(k,z)\, k^2\right]^{-1} \tag{44}$$

where the generalized self-diffusion function $\tilde{D}_s(k,z)$ is the memory function of G. There is again a generalized Stokes-Einstein relation (compare with (18)) for the case of self-diffusion

$$\tilde{D}_s(k,z) = \frac{c_o^2}{z + \frac{1}{m}\tilde{\zeta}_s(k,z)} \quad ; \quad c_o^2 = k_B T/m \tag{45}$$

We only quote the long-wavelength limit of $\tilde{\zeta}_s(k,z)$, since that is sufficient for all further results

$$\tilde{\zeta}_s(0,z) = \zeta_o + \beta < F_{1z}(z - \hat{Q}_1 \hat{\Omega}\, \hat{Q}_1)^{-1} F_{1z} >_o \tag{46}$$

Here, F_{1z} is the z component of the force exerted on the tagged particle by all other macroions. $\hat{Q}$ is a projector perpendicular to $\hat{c}_1(\underline{k}) = \exp(-i\underline{k}\cdot\underline{r}_1)$ and $\hat{j}_1(\underline{k}) = \underline{p}_1 c_1(\underline{k})/m$. From (43) to (45) the velocity auto-correlation function is

$$\tilde{Z}(z) = \tilde{D}_s(0,z) = \frac{c_o^2}{z + \frac{1}{m}\tilde{\zeta}_s(0,z)} \quad , \tag{47}$$

so that (46) is essentially the memory function of $\tilde{Z}(z)$.
The self-diffusion coefficient is given by

$$D_s = \int_0^\infty dt\; Z(t) = \tilde{Z}(0) = \tilde{D}_s(0,0). \tag{48}$$

The expression (46) has again been evaluated in a mode-mode coupling approximation. The result is

$$\tilde{\zeta}_s(0,0) = \zeta_o + \frac{k_B T\, c}{3(2\pi)^3} \int d^3k' \left\{ \frac{k_z'\, h(\underline{k}')}{S(k')} \right\}^2 \int_0^\infty dt\; G(\underline{k}',t)\, S(\underline{k}',t) \tag{49}$$

The time integral is performed with mean-field expressions for the two correlation functions $G(k,t) = \exp(-D_o k^2 t)$ and $S(k,t) = S^{MF}(k,t)$. The resulting self-diffusion coefficient is[16]

$$D_s = D_o \left[1 + \frac{c}{6\pi^2} \int_0^\infty dk' \frac{h^2(k')}{1 + S(k')}\right]^{-1} \tag{50}$$

so that D_s is completely reduced to a knowledge of the static structure factor $S(\underline{k})$. The self-diffusion coefficient is always reduced due to the interactions, $D_s \leq D_o$. For a Debye-Hückel system

$$S(k) = 1 - c\, h(k) = 1 - \frac{\kappa_B^2}{\kappa_{DH}^2 + k^2} \tag{51}$$

where

$$\kappa_B^2 = 4\pi\beta c\, Q^2/\varepsilon \quad ; \quad \kappa_{DH}^2 = \kappa_B^2 + 4\pi\beta \sum_\alpha c_\alpha q_\alpha^2/\varepsilon \, . \tag{52}$$

Here, c_α is the concentration of the counter ions of species α carrying the charge q_α. Using (51) in (50) leads to

$$D_s = D_o \left[1 + \frac{Q^2 \kappa_{DH}}{3\varepsilon k_B T} \left(1 - \left[1 - \frac{\kappa_B^2}{2\kappa_{DH}^2}\right]^{1/2}\right)\right]^{-1}$$
$$\approx D_o \left[1 - \frac{Q^2 \kappa_{DH}}{3\varepsilon k_B T} \left(1 - \left[1 - \frac{\kappa_B^2}{2\kappa_{DH}^2}\right]^{1/2}\right)\right] \tag{53}$$

The last line can also be obtained from a result derived by Onsager[17] for a mixture of ions, if one chooses the friction coefficient of one species of ions much larger than all other friction coefficients.

For the system of polystyrene spheres (51) is not applicable. Therefore, the structure factor calculated in Sect. II according the scheme developed by Hansen and Hayter has been used in (50), and the result for D_s as a function of concentration is shown in Fig. 3.

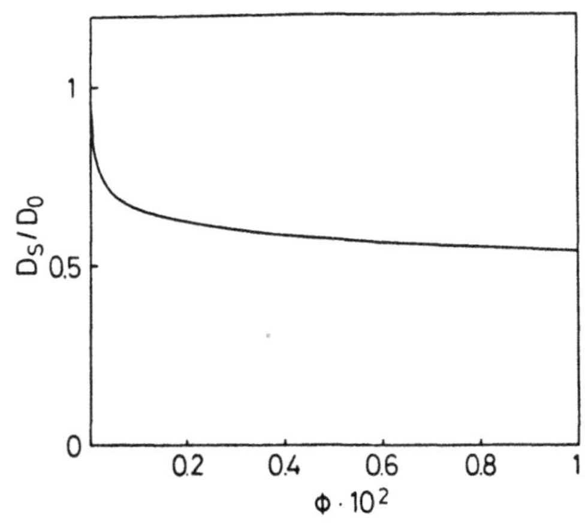

Fig. 3. Theoretical result for the concentration dependence of the self-diffusion coefficient of polystyrene spheres with $d = 900$ Å, $\psi_o = 73$ mV, $\kappa^{-1} = 5000$ Å.

Finally, we make a few remarks concerning the long-time behavior of the velocity autocorrelation function. Treating (46) in mode-coupling approximation it is found that $\zeta_s(0,t)$ behaves as $t^{-5/2}$ for $t \to \infty$. Using instead a weak-coupling Debye Hückel static structure factor, $S(k) = k^2/(k^2+\kappa^2)$, we find $\zeta_s(0,t) \sim t^{-3/2} \exp(-\nu_p t)$, where $\nu_p = \tau_B \omega_p^2 = D_o \kappa^2$ is the overdamped plasma frequency of a one-component plasma (OCP), ω_p being the usual plasma frequency[18]. Since $\zeta_s(0,t)$ is the memory function of $Z(t)$, the long-time tail of $Z(t)$ is given by the one of $\zeta_s(0,t)$ up to a negative constant factor[19]. It is interesting to compare our result for the overdamped Debye-Hückel system with the long-time tail for the ordinary OCP, which is[20] $t^{-3/2} \cos(\omega_p t)$. Since in our system there are no plasma oscillations, ω_p is replaced by ν_p and the cosine function becomes an exponential decay. We would also like to mention that an earlier treatment[21] of the strongly interacting polystyrene sphere system on the basis of the Smoluchowski equation gave a velocity autocorrelation function which could well be represented by $t^{-3/2}\exp(-\alpha t)$ for a substantial time interval .

VII. Conclusions

The dynamical properties of an overdamped system of charged hard spheres has been developed on the basis of the Fokker-Planck equation. The main difference to the theory of simple liquids is the presence of the friction and as a consequence of it a smaller set of conserved quantities. The dynamic structure factor is expressed by generalized diffusion and viscosity functions which are closely connected with the memory functions of the concentration autocorrelation. These functions have been calculated by a mode-coupling approximation. Using the static structure factor as an input, good agreement with the results of quasi-elastic light scattering experiments is obtained. A similar treatment of one-particle properties leads to a prediction for the concentration dependence of the self-diffusion coefficient.

Acknowledgement

We would like to thank Dr. J. Hayter, ILL Grenoble, and Prof. J.P. Hansen, Paris, for providing us with their computer program to calculate S(k). One of the authors (W. H.) acknowledges the hospitality during a stay at the ILL, where part of this work has been done.

References

1. H.Z. Cummins and E.R. Pike, eds., Photon Correlation and Light Beating Spectroscopy (Plenum Press, New York, 1974).
2. A recent comprehensive review is given by P.N. Pusey and R.J.A. Tough -n: Dynamics of Light Scattering and Velocimetry: "Application of Photon Correlation Spectroscopy", ed. by R. Pecora, Plenum, New York, to be published. See also J.M. Schurr, Crit. Revs. in Biochemistry 4, 371 (1977).
3. J.C. Brown, P.N.Pusey, J.W. Goodwin, and R.H. Ottewill, J. Phys. A8, 664 (1975).
4. P.S. Dahlberg, A. Bøe, K.A. Strand, and T. Sikkeland, J. Chem. Phys. 69,5473(1978).
5. F. Grüner and W. Lehmann, J. Phys. A 12, L-303 (1979).
6. E.J.W. Verwey and J.Th.G. Overbeek, Theory of the Stability of Lyophobic Colloids (Elsevier, 1948).
7. J.B. Hayter and J. Penfold, Mol. Phys. 42, 109 (1981).
8. J.P. Hansen and J.B. Hayter, Mol. Phys., to be published.
9. H. Minoo, C. Deutsch, and J.P. Hansen, J. Physique Lettres 38, L-191 (1977).
10. J.M. Deutch and I. Oppenheim, J. Chem. Phys. 54, 3547 (1971).
11. T.J. Murphy and J.L. Aguirre, J. Chem. Phys. 57, 2098 (1972).
12. H.M. Fijnaut, C. Pathmamanoharan, E.A. Nieuwenhuis, and A. Vrij, Chem. Phys. Lett. 59, 351 (1978).
13. W. Hess and R. Klein, to be published.
14. W. Hess, J. Phys. A 14, L-145 (1981).
15. B.J. Ackerson, J. Chem. Phys. 69, 684 (1978); W. Dieterich and I. Peschel, Physica 95A, 208 (1979).
16. W. Hess and R. Klein, to be published.
17. L. Onsager, Ann. N.Y. Acad. Sci. 46, 241 (1945).
18. An overdamped OCP was treated on the basis of the Smoluchowski equation by W. Hess and R. Klein, J. Phys. A 13, L-5 (1980).
19. N. Corngold, Phys. Rev. A6, 1570 (1972).
20. H. Gould and G.F. Mazenko, Phys. Rev. A15, 1274 (1977).
21. W. Hess and R. Klein, Physica 105A, 552 (1981).

NUCLEIC ACIDS AND POLYELECTROLYTE THEORY*

M. Guéron, Groupe de Biophysique
Ecole Polytechnique, 91128 Palaiseau

1. TRANSFER NUCLEIC ACIDS AS MODEL POLYELECTROLYTES

1.1. Structure of nucleic acids

Most synthetic polyelectrolytes are polydisperse, and are studied in conditions where they are flexible and interacting. By contrast, there exists nucleic acids of precise molecular weight and chemical composition, which may assume a unique structure in appropriate conditions.

Nucleic acids come in many types, and they can exhibit different structures. For instance deoxyribonucleic acid (DNA) is a double-stranded polymer. In solution it generally exhibits the famous "B" form, a right-handed double helix with hydrogen-bonded base pairs perpendicular to the helix axis. However, the "Z" form, a left-handed helix, has been demonstrated recently. In association with basic proteins (the histones), DNA forms nucleosomes, i.e. structures in which a 60nm length of double helix is wound in two turns of a super helix, as shown by crystallography. These different structures are most probably involved in the regulation of biological activities (replication, transcription) of DNA.

Structurally, the best known species of nucleic acids are the transfer ribonucleic acids (tRNA), which act as specific carriers of amino-acids in the process of protein synthesis. For instance

---

* This work is dedicated to the Russian scientist Victor BRAILOVSKI.

yeast $tRNA^{Phe}$, the carrier for phenylalanine from yeast, is a linear polymer 76 nucleotides long (molecular weight ca. 25000 daltons). In the so-called native structure, the chain folds upon itself by internucleotide associations which involve mainly stacking and hydrogen bonding. This

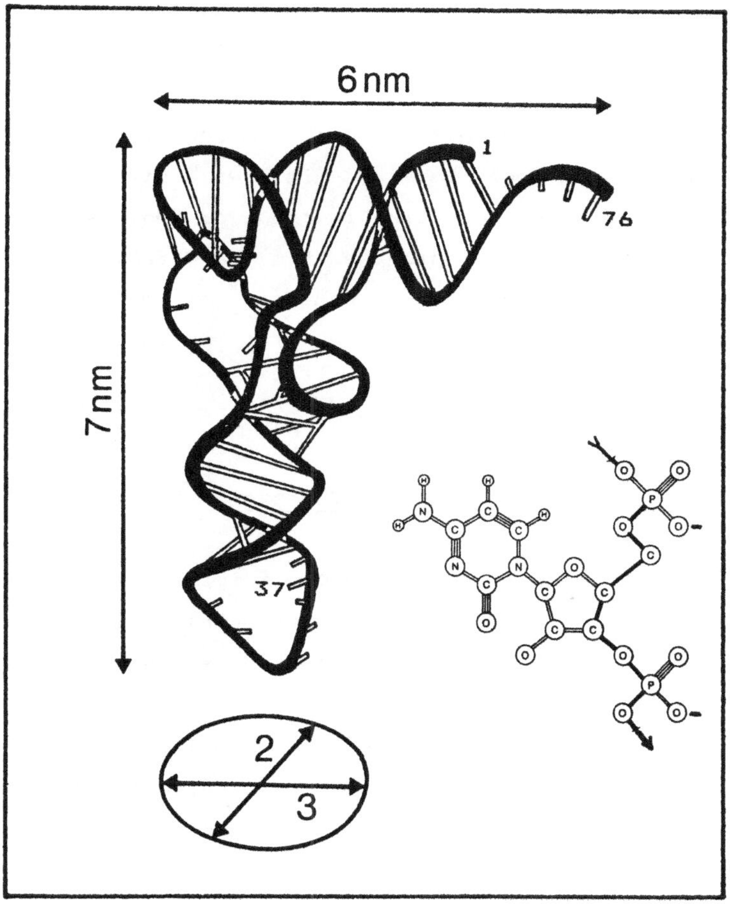

Figure 1a) The three-dimensional structure of yeast $tRNA^{Phe}$. The black line represents the phosphate-ribose backbone, and the rungs show bases and base pairs. There are 75 singly charged phosphodiesters plus one terminal phosphoester. For clarity a skeletal representation has been chosen, but the molecule is in fact highly compact, all parts being in Van der Waals contact. The charged phosphates are mostly on the outside.

b) A fragment of the polynucleotide chain, showing phosphate, ribose and base. The bases are of 4 types: guanine, cytosine, adenine and uracil. Some are chemically modified. The base shown here is a cytosine.

results in a very compact, L-shaped structure, which has been determined to about 0.25nm resolution by X-ray crystallography[1] (fig.1). There are 76 singly charged phosphate groups, nearly all on the outside of the molecule, for an average surface area A per electronic charge, of roughly 1 nm^2. This is quite a large charge concentration, and the mutual electrostatic repulsion is a significant factor in the determination of the structure.[2]

In fact, the native conformation of nucleic acids in general is achieved only in high salt. At a given temperature, they "melt" if the salt concentration is reduced; conversely, the melting temperature increases in high salt. The stabilizing effect of divalent ions is much larger than that of monovalent ions. The structure is altered if the dielectric constant of the solvent is reduced (alcohol-water mixtures).

1.2. Electrostatics and biological function

In what ways is the large electrical charge involved in the course of biological processes ? No precise answer is known at present, but the following suggestions can be made.

The biological functions of nucleic acids nearly always involve association with proteins. The association may lead to structural changes of either partner. It is often highly specific.

Some proteins (e.g. endonucleases) distinguish between single- and double-stranded nucleic acids: this may involve the difference in the number of charges per unit length. Other proteins (RNA polymerases, DNA polymerases) migrate along the DNA chain without becoming detached: electrostatic interactions may be an agent of such restricted motion.

The internal matter of proteins is weakly polarizable: the dielectric constant is close to one. It is therefore energetically costly to bring a charged phosphate in the neighborhood of a protein, unless it is neutralized by an adjacent positively charged group

($-NH_3^+$), or unless it associates with polarizable surface groups (-OH), or unless bridging counterions or water molecules are involved. Either way, it seems that the presence of charged phosphate groups may restrict the possible geometry of the association and thereby contribute to stereospecificity. Regarding conformational changes, the electrostatic repulsion contributes an internal stress that can be modulated by varying the charges and/or the dielectric constant in the neighborhood. Such mechanisms may be exploited, for example, by proteins that open double-stranded helices of DNA.

1.3. Physical chemistry of transfer nucleic acids and their interactions with ions

The phenomena alluded to above raise complex problems, involving equilibrium structure, alternative conformations, intermolecular interactions etc. Their understanding requires analysis of a simpler, yet relevant system. As mentioned above, a transfer RNA provides a system of known composition and defined structure. Numerous spectroscopic studies have been applied to the question of its interaction with counterions.

For instance, quenching of the fluorescence of a centrally located group by manganese shows that when divalent ions are added, the first one goes at least part time to this area of the molecule. In constrast, fluorescence of a nucleotide near the anticodon tip of the molecule is enhanced only after addition of about 5 Mn, in low-salt solutions. The discrimination between center and anticodon tip is reduced at higher NaCl concentration[3](fig.2).

Perturbation of the ^{31}P nmr spectrum by manganese and cobalt provides structural and kinetic probes.[4-6] Addition of manganese broadens and relaxes the phosphorus resonances, which are also shifted by cobalt. The residence time and average phosphorus-to-manganese distance can be determined. The interpretation of the experiments in terms of binding sites of the divalent ions is limited because the ^{31}P spectrum is presently not assigned, except for the terminal phosphate. Nevertheless, a considerable amount of information is obtained:

a) no phosphate remains unaffected by the divalent ions, which therefore migrate over the entire molecule. No phosphate is visited more than 7 times more often, or less than 7 times less than the average phosphate (fig.3).

b) the sticking time is in the range of 10^{-7} second at room temperature. This is much longer than expected if the ions diffuse freely, but much shorter than could be envisioned if the ions were chelated in a site such as is formed in metallo-enzymes. On the contrary, this time is quite comparable to the sticking time found

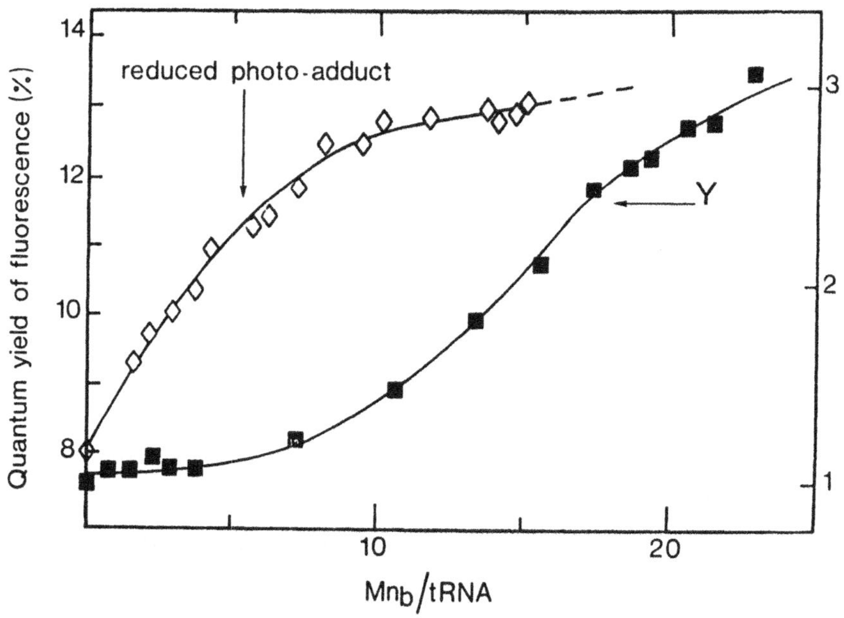

Figure 2 When added to a solution of tRNA in 5 mM NaCl and 10 mM triethanolamine chloride buffer, divalent ions such as manganese induce the formation of the native structure. This is followed in E.Coli $tRNA^{fMet}$ by the increased fluorescence of a base derivative created photochemically, which is located in the center of the molecule (◇). In yeast $tRNA^{Phe}$, the Y base (n°37) is located at the anticodon tip of the molecule. In both cases the fluorescence is enhanced when conformation changes result in protection of the bases from water. Such changes are seen to occur in the center part of the molecule as soon as the first divalent ions are added, whereas the Y base fluorescence changes only after binding of 5 Mn per tRNA(■).

for simple ligand binding to manganese. And indeed similar experiments with cobalt yield sticking times in the range of 10^{-5} second, once again in line with the time for binding o- simple ligands.

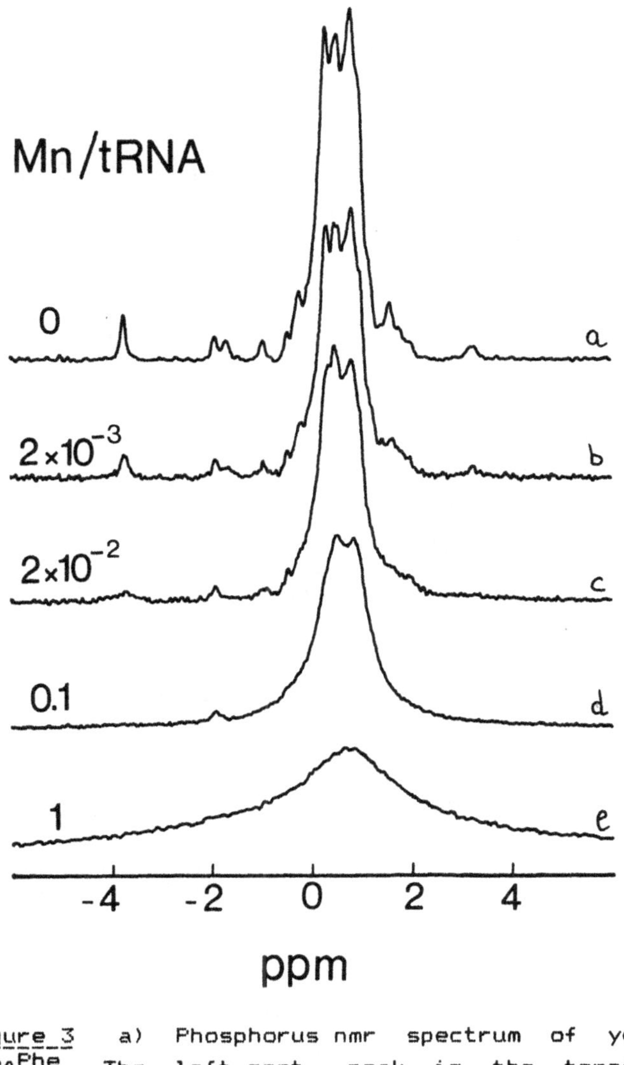

Figure 3 a) Phosphorus nmr spectrum of yeast $tRNA^{Phe}$. The left-most peak is the terminal phosphate. The other isolated peaks correspond to phosphates shifted from the main peak by structural effects (on melting the structure, the spectrum collapses to a unique position, at 0.4ppm).

b) to e) Addition of manganese broadens the phosphate resonances. Since all phosphates are affected even at low manganese concentrations, the ion must migrate over the entire molecule. The temperature is 30°C; the solution contains 150 mM NaCl; pH=7.4.

c) the average phosphate-to-manganese distance is found to be a function of temperature, suggesting a temperature-dependent equilibrium between inner and outer-sphere ligation (fig.4). This has prompted us to search for such an equilibrium in mononucleotides as a model. In the cobalt-adenosine monophosphate complex, we have indeed demonstrated directly the existence of two varieties, using both proton and phosphorus nmr.[7]

The experiments just described are examples of detailed and local investigations of a polyelectrolyte which is not only monodisperse but perfectly defined in its composition, and endowed with a specific structure. Many other physico-chemical studies of transfer RNAs are being done. Some of the potentially most

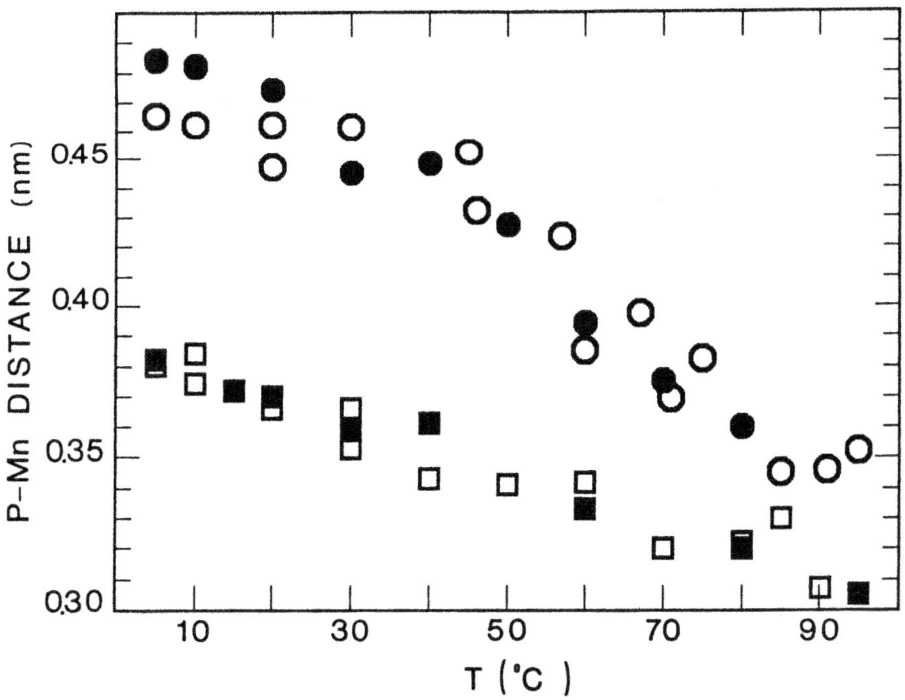

Figure 4 Longitudinal relaxation of phosphorus by manganese is analyzed to yield an average phosphorus-to-manganese distance, in various polynucleotides: tRNA (O); calf thymus DNA, 30 base-pair long (●); single- stranded polynucleotides A (■) and U (□). Average distances differ for double-stranded and single-stranded species. The temperature dependence suggests an equilibrium between different complexes. The direct (inner-sphere) binding distance is 0.28 nm.

informative are based on the proton nmr spectrum, in which each base pair provides a single line due to an exchangeable proton (fig.5). The assignment of the twenty-odd peaks is advancing rapidly and will provide as many punctual probes of molecular structure and dynamics.[8] Lastly one should not forget an important advantage of transfer RNAs as physico-chemical models: there are a larger number of variants, roughly one for each codon, and their conposition depends on the species from which they originate.

For all these reasons, it would seem that transfer RNAs could be taken as a reference for polyelectrolyte concepts, studies, and theories.

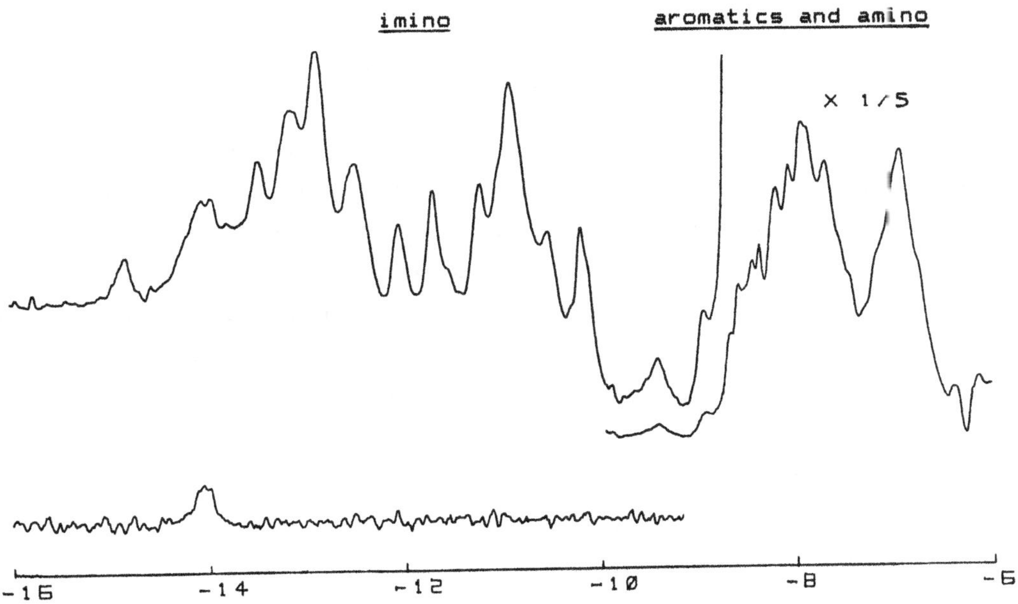

Figure 5 a) The low-field part of the proton nmr spectrum of 1mM yeast $tRNA^{Asp}$, showing the peaks of imino protons involved in hydrogen bonds. The spectrum is obtained in the presence of H_2O whose resonance (at -4.8ppm), corresponding to a concentration of 100 M/l, is suppressed by a special pulse sequence.

b) In D_2O, all imino protons exchange, and their spectrum disappears, at rates generally in the range of seconds to hours. One proton exchanges very slowly, demonstrating that the corresponding base pair hardly ever opens up (the lifetime is 4 days at 20°C, with 15 Mg/tRNA). This base - pair is now identified with near-certainty.

2. A THEORETICAL FRAMEWORK

2.1. The elements

Polyelectrolytes are highly charged macromolecules. They therefore create an electrostatic potential around them. Generally speaking, the potential may modify the structure and motion of the macromolecule, the orientation, polarizability and dielectric constant of the solvent, the distribution of counterions and possibly their state of hydration, not to speak of interactions between the polyelectrolytes themselves. The extreme complexity of the problem implies that drastic simplifications are necessary.

The properties accessible to experiment are mainly of three types:

a) thermodynamic properties, such as the counterion or co-ion activities. They are insensitive to the ionic distribution and interactions near the polyelectrolyte surface. In many cases they do not discriminate effectively between different theories[9] (table I).

b) properties of the counterions near the polyelectrolyte surface. They depend on the distribution of the counterions and on their molecular interactions with charged or uncharged groups of the polyelectrolyte.

c) transport properties: they involve features from both a) and b). Their analysis is generally difficult.

The region near the polyelectrolyte is of greatest interest. This is where polyelectrolyte effects originate. It is a key area for the theory. It is now experimentally accessible via powerful spectroscopic methods.

Our theoretical approach is as follows:

a) We start with a polyelectrolyte of given shape. The solvent is continuous and the ions punctual. The solvent dielectric constant is unaffected by the electric field. The polyelectrolyte charge is distributed continuously on its surface.

b) We take a mean field approach and compute the ccunterion distribution.

c) We next deal with the chemical interactions of the counterions with the polyelectrolyte groups. These interactions are considered to take place exactly as with identical but monomeric groups. We assume that the only electrical effect of the polyelectrolyte is to alter the ionic concentrations. We neglect such effects as solvent polarization.

For instance, interactions of magnesium with tRNA are considered as equivalent to interactions of magnesium at the appropriate concentration (CIV:concentration in the immediate vicinity of the polyelectrolyte) with a collection of phosphodiester nucleotides. The usual

X	γ^m	γ^c	γ^{PB}
1.2	0.47	0.50	0.49
1.4	0.36	0.46	0.47
2.5	0.35	0.34	0.35
3.1	0.29	0.30	0.31
3.1	0.25	0.30	0.31
6.1	0.21	0.21	0.21
6.2	0.20	0.20	0.21
7.7	0.16	0.18	0.19
12.2	0.14	0.15	0.15
15.4	0.13	0.13	0.14
∞	0.075±0.05	0.072	0.083

Table I Counterion activity coefficients in DNA/$MgCl_2$, for various polymer concentrations.

γ^m is the measured value (Lyons, J.W. and Kotin, L. (1965) J. Am. Chem. Soc., 87, 1670); γ^c is the value computed by condensation theory,[9] and γ^{PB} is the Poisson-Boltzmann value.

The condensation and Poisson-Boltzmann values are practically the same, showing that the prediction of counterion activity is not a good discriminator between theories.

(X is the phosphate to chloride ratio.)

experimental and theoretical criteria are used to analyze which groups (phosphate, base nitrogen, ribose-OH etc.) coordinate to the ion, whether coordination occurs via water or not, what rate constants govern these interactions, and in what proportion the ions are coordinated. The latter property is governed by the mass-action law, where the CIV plays the role of the free ion concentration.[10]

This theoretical framework is not original. It has been used by electrochemists for three quarters of a century, and is associated with the names of Poisson, Boltzmann, Gouy, Chapman and Stern. The same inspiration appears in polyelectrolyte studies such as those of the Katchalsky group and others.

A well-known blight of Poisson-Boltzmann theory is that it is inconsistent in the high-field, non-linear case:[12] the mean quantities (potential, concentrations, etc.) of course exist, but they cannot be derived from equations involving only mean quantities. Higher-order quantities such as ion-ion correlations must be taken into account. This is accomplished, in principle, by more elaborate theories (e.g. cluster-theory[13]) which however give rise to serious computational difficulties. Luckily, there is some evidence that in the case of highly-charged polyelectrolytes, Poisson-Boltzmann inconsistency is not too severe.[14] Furthermore, the theory can be corrected to take account of inter-ion correlations.[9]

The first question that we investigated with the help of the Poisson-Boltzmann theory was the influence of polyelectrolyte shape, since we had observed that divalent ion binding to tRNA and DNA where quite similar.[15] Numerical solutions for planar, cylindrical and spherical charged surfaces were indeed found to be comparable in the vicinity of the interface.[10] This lead us to propose a description of the Poisson-Boltzmann solution near the planar interface, and its extension to curved surfaces.[16] More recently we have introduced the concept of a scaling length in order to predict the domain of validity of the solution for curved surfaces.[17] The principal results of the theory are given below. They are obtained in the approximation of point ions. (Poisson-Boltzmann treatments of finite size corrections have been presented[18,19]).

2.2. Results: sum-rule for CIVs, scaling length, shape, site-binding.

We consider a polyelectrolyte surface in a solution containing ions of charge $z_i e$, in concentration n_i. The specific area A is defined as e/σ where σ is the surface charge density of the polyelectrolyte. The principal results are as follows (the electrostatic potential is designated Ψ):

a) There is a sum-rule for the plane[16,20]:

$$\sum (CIV_i - n_i) = 2\pi l_B/A^2$$

where l_B is the Bjerrum length: $l_B = e^2/(4\pi\epsilon_0 Dkt)$, t being the temperature. This shows that the sum of excess concentrations at the

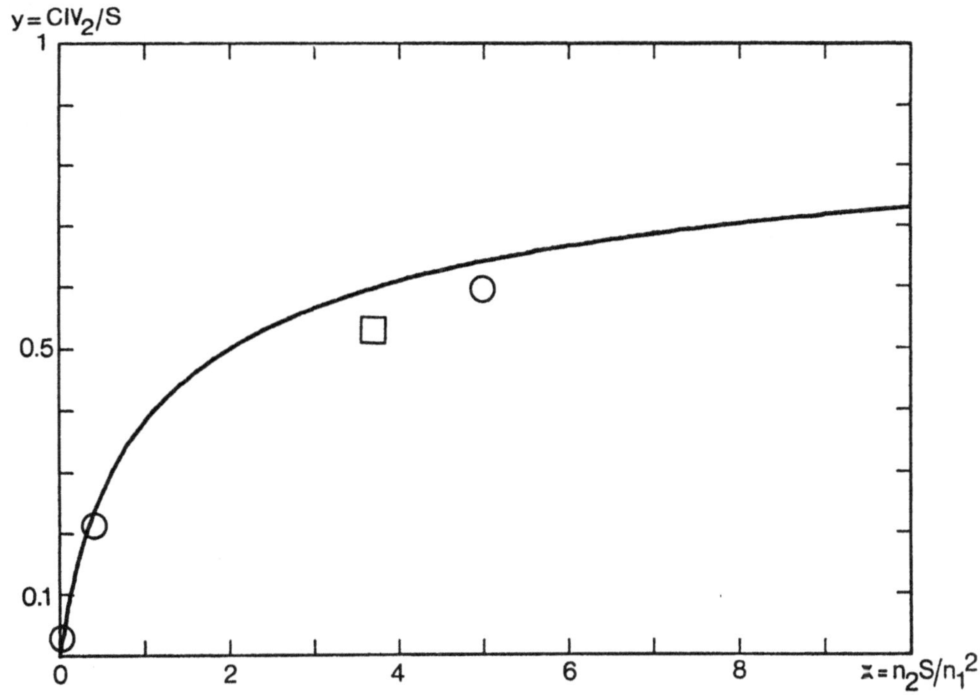

Figure 6 Universal graph for determining the counterion concentrations at the surface of a highly charged plane, in a solution containing monovalent and divalent cations. S is the (constant) sum $CIV_1 + CIV_2$, and is equal to $2\pi l_B/A^2$. The symbols corresponds to a cylinder having $\xi = 5$, in solutions of different concentrations: the planar and cylindrical cases are very similar.

interface is independent of the ion charges and of their concentrations at large. Furthermore, we have $CIV_i = n_i \exp - z_i\omega$ where ω is the value at the surface of $\varphi = \psi e/kt$. Hence the sum-rule provides an algebraic equation for $\exp\omega$. Simple computations or graphs then enable one to determine the CIV of each ion and various other parameters, for instance the number N_z of accumulated divalent ions in the case of a mixture of mono- and z-valent counterions (fig.6). A straightforward result is that N_z is proportional to n_1^{-z}, at low concentrations of the z-valent species. This result is significant for mixtures of mono- and divalent ions, as well as for the binding of charged proteins to DNA.

b) Let us define a scaling length l_e for the variation of the electric field near the surface: l_e is the value at the surface of $(d\varphi/dx)/(d^2\varphi/dx^2)$. It is then found[17] that for a plane in pure salt ($z_i = \pm z$), the charge concentration at the interface, its variation near the interface, and the integrated charge as well, can all be expressed as a function of l_e (table II). This provides a unified description for the low-potential (Debye-Hückel) and high-potential (non linear) regimes. In the former regime, l_e is equal to the Debye length λ, defined by $\lambda^{-2} = \sum 4\pi z_i^2 \times n_i$; in the latter l_e becomes independent of the solution concentration: $l_e \rightarrow T = A/(2\pi z \times l_B)$, as shown in fig.7.

		linear regime	non-linear regime
(1) scaling length	l_e	$\lambda = (4\pi l_B \sum n_i z_i^2)^{-1/2}$	$A/(2\pi l_B z)$
(2) integrated charge up to distance l_e	$A\int_0^{l_e} \rho(x)dx$	$(e-1)/e$	$1/2$
(3) charge concentration at the interface	$\rho(0)$	$1/(l_e A)$	$1/(l_e A) = 2\pi l_B\, z/A^2$
(4) charge concentration at distance l_e	$\rho(l_e)$	$\rho(0) \times (1/e)$	$\rho(0) \times (1/4)$

Table II. Description of the charge distribution near a charged plane. Using the scaling length l_e, similar expressions obtain for the parameters of the distribution, in the linear and non-linear regimes.

The introduction of l_e provides a plausible criterion for the extension of the planar formulae to curved surfaces. Such an extension should be valid for $R_c > l_e$, where R_c is the curvature radius. In the linear regime, this gives the trivial condition $R_c > \lambda$, whereas in the non-linear regime, the relation $R_c > T$ leads to $2z\xi > 1$, where ξ is the linear charge density, originally defined for a cylinder[21] and extended by us to other shapes.[10] This suggests that the plane is a good starting point for the study of highly charged polyelectrolytes.

A perturbation-empirical approach suggested by such considerations and verified by numerical computations provides a formula for the surface potential φ_s of a curved surface, by reference to the surface potential ω of a plane having the same surface charge density:

$$\varphi_s = \omega - (2/z')/\{\xi z' - 0.45 + 1.8[\varepsilon/(\xi z')]^{1/2} + 3\varepsilon/(\xi z')\}$$

where $\varepsilon = R_c/\lambda$; in pure salt z' is equal to the counterion valency; in mixed salt, it is an equivalent valency. The (extensive) range of validity of this formula has been discussed previously.[16]

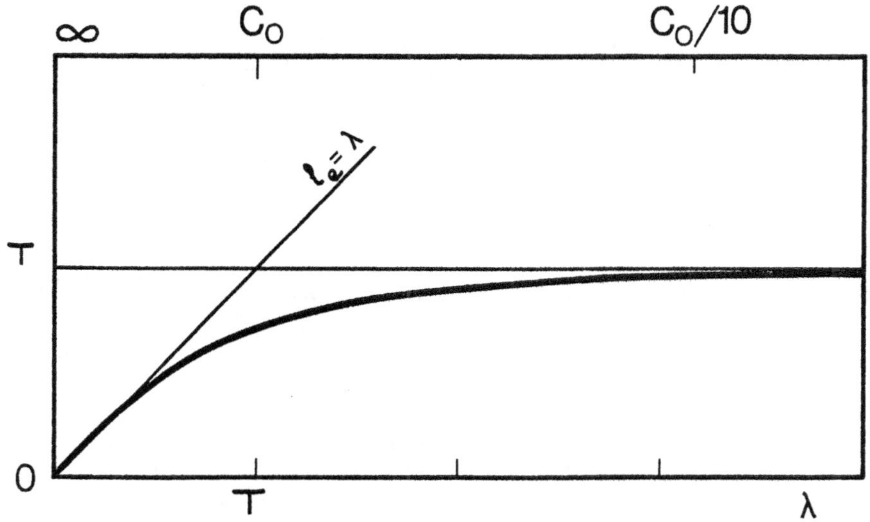

Figure 7 The scaling length for a charged plane in solutions of pure salt (the valency is z). The concentration C is represented by the Debye length λ, in abcissa. Left of $\lambda = T$, the concentrations are high: this is the linear case, and $l_e = \lambda$. To the right, the concentrations are low: non-linear case, $l_e = T$.

c) When site-binding is taken into account, the surface charge density of the polyelectrolyte is decreased, and this in turn reduces the CIV. We have studied the extent of binding in a pure salt solution as a function of the equilibrium constant K and the salt concentration.[22] The extent of binding is weakly dependent on K, and even less on the salt concentration. Due to the large value of the CIV for highly charged surfaces, binding is significant even for low values of K (fig.8). Thus the model of a macromolecule carrying a large net charge density is often unrealistic.

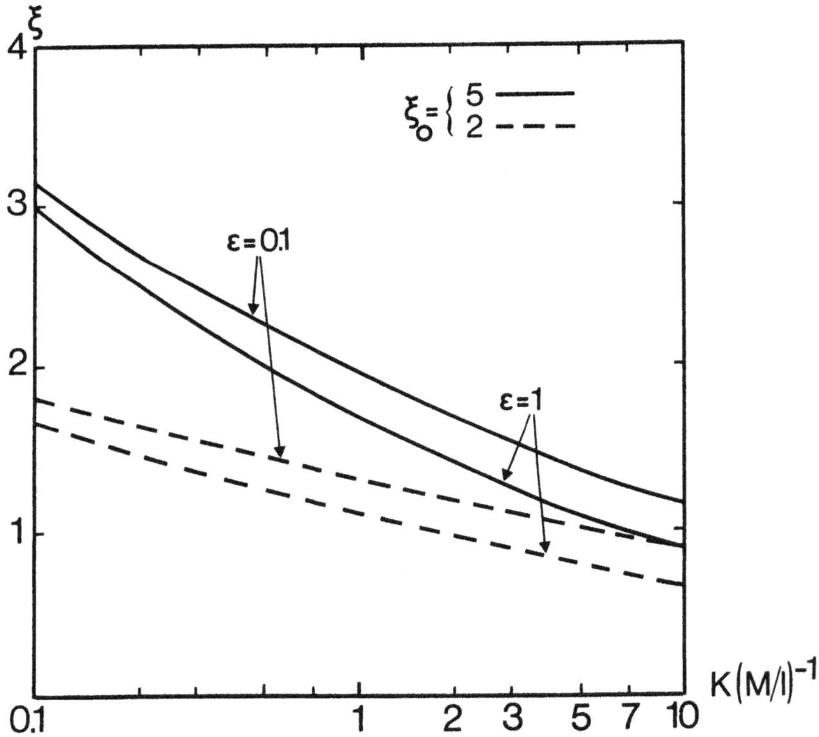

Figure 8 Reduction of the linear charge ξ of a polyelectrolyte cylinder by site-binding of counterions. The structural charge density $\xi_o = 5$ is reduced to values between 1 and 3 (full lines), depending mainly on the binding constant K and only slightly on the salt concentration. Values of K in the range of 10 and 0.1 $(M/l)^{-1}$ are expected for the interaction of phosphates with magnesium and sodium, respectively. Values of $\mathcal{E}$ = 1 and 0.1 correspond to 100 and 1 mM NaCl when the cylinder diameter is 2nm. The dotted lines correspond to a cylinder whose structural charge is ξ_o = 2.

2.3. Concluding remarks

As just described, the Poisson-Boltzmann theory provides an excellent tool for the investigation of biological polyelectrolytes. It has been applied to the interaction of tRNA and divalent ions, in particular to the interpretation of the binding curves in solutions containing mono- and divalent ions. Manganese epr and phosphorus or proton nmr studies are currently interpreted along the same lines. The flexibility of tRNA, its melting behavior, its interactions with proteins should also be approached similarly in the future. The same methods apply to other nucleic acids (DNA) and other polyelectrolytes as well.

A prerequisite for these analyses has been an understanding of the effect of polyelectrolyte shape on the electrostatic problem in the high-field regime.

3. THE CONDENSATION PHENOMENON

3.1. On condensation theory

In the first part of this work, we have discussed the role of electrostatic factors in the structure and function of nucleic acids, emphasizing the importance of the region close to the polyelectrolyte. We have presented some experimental methods for probing the structure of the polyelectrolyte, the properties of nearby ions, and the ion-polyelectrolyte interactions.

In the second part, we presented a theoretical framework for dealing with this problem. This consists essentially in the observation that for highly charged surfaces, shape-dependence is weak, so that the ion distribution near a polyelectrolyte is similar to that near a charged plane, a case which has been extensively studied. The theory then provides simple, detailed, quantitative descriptions of the electrostatic properties. Its limitations are understood. Extensions are straightforward in principle. For instance, computation of the free energy for a given shape leads to investigation of electrostatic effects on shape, etc.

It would therefore seem that there is no room for an alternate theory, unless it is significantly better in its assumptions or formalism, or in its agreement with experiment. Since an alternate theory (condensation theory[23]) is frequently evoked, one may wonder whether it satisfies the above expectations.

Condensation theory has the significant property that it is based on electrostatics and thermodynamics, and is a mean-field theory. Such a basis, if carried through coherently, leads mathematically to Poisson-Boltzmann theory, to which condensation theory should therefore be equivalent. Nevertheless condensation theory leads to a different distribution of the counterions. This can only be due to error in its derivation. We have pointed out[10] such an error, which takes the guise of an unstated assumption in the demonstration of the existence of a condensation radius[23]. The theory has also been criticized on other grounds[24].

In the absence of good theoretical arguments, is there an experimental backing for condensation theory? The answer here is necessarily more complex since it involves first the experimental data and their interpretation, second the theoretical constructs and third the comparison between them. And if agreement is obtained, does it supports the theory rather than a competing one?

In what follows, we examine recent experiments which have been interpreted in favor of condensation theory. The experiments themselves will not be discussed here, but we shall show that the comparison with condensation theory is confused, in particular regarding the question of ion binding, so that the experiments do not in fact support the theory.

3.2. Theory and experiment

As a first example, consider a study of ^{23}Na transverse relaxation in the presence of DNA[25]. It was found that the number of Na ions interacting with DNA is independent of the sodium concentration at large. This was claimed to support condensation theory, because alternative interpretations, involving association according to the mass-action law, would have, in the authors' view, implied a strong effect of the salt concentration at large. This is incorrect, for it is the CIV, not the concentration at large which corresponds to "free ion concentration" in the mass-action law. And indeed the CIV is only weakly sensitive to the concentration at large.

Another example is a series of three interesting articles dealing with manganese and cobalt binding to DNA fragments, as studied by NMR relaxation of phosphorus and of water protons[26]. In the first paper, the conclusion was drawn that only 5 to 20% of the counterions were inner-sphere coordinated to phosphate. This was said to be much less than expected according to an analysis of binding along the lines of the present work. It was argued that the weak amount of inner-sphere binding to phosphate agrees qualitatively with condensation theory, which states that binding does not occur.

However a consideration of bound ions must take account not only of inner-sphere binding to phosphate, but of all types of binding : inner or outer-sphere, to phosphates or other groups. And indeed, this is the definition of binding proposed in the two other papers of the series : the second paper provides evidence that of all divalent ions close to DNA, a large proportion is indeed bound. This proportion is as large as 95 % for low manganese-to-phosphate ratios, according to the authors' table V. (Because of a plotting error, their figure 6 does not show this). It is this fraction of 95 % (rather than 5 to 20 %) that should be compared to the no-binding proposition of condensation theory. Quoting from ref.23 (page 215) :

"...in particular, Mg^{2+} (and probably many other divalent metal cations) binds to polynucleotides in the delocalized mode. The immediate environment of a bound Mg^{2+} ion is very nearly the same as in free solution ; the bound state is characterized by essentially complete hydration and free translational and rotational mobility."

Where is the agreement between experiment and theory ?

In the third paper, the number of bound ions is compared to predictions made on the basis of condensation and Poisson-Boltzmann theories. For the comparison to condensation theory, the authors assimilate the experimentally bound ions to the condensed ions of the theory, whereas a key point of condensation theory is that condensed ions are not bound. (Possible reasons for this have been proposed earlier [10]).

Unlike condensation theory, Poisson-Boltzmann theory is easily extended to incorporate binding, as discussed above. However, in comparing the experiments to Poisson-Boltzmann theory, the authors have not carried out such a computation. Instead they improperly assimilate the experimentally bound ions to the accumulated, non-bound ions of the theory.

Despite the questions just raised, it is interesting to consider the authors' conclusions, namely that condensation and Poisson-Boltzmann theories lead to similar predictions, both being more or less in agreement with experiment. These statements, with which we disagree, might lead the reader to believe that there is not much difference between the two theories. The explanation is quite different and lies in two observations.

First, what is computed is the number of ions close to the polyelectrolyte. This is the complement to the number of screening ions at large, a number which determines the activity of the counterions and is not sensitive to the theoretical model nor to the extent of binding[9]. Hence the apparent agreement between both theories regarding the number of ions close to the polyelectrolyte.

Second, why does this number more or less agree with the number of experimentally bound ions, whereas one would expect it to correspond to the total number of ions, bound and unbound, close to the polyelectrolyte? The reason seems to be simply that most of the ions are bound, according to the second paper, so that these two numbers are about the same.

Support for condensation theory, which negates binding, is thus drawn from experiments in which the ions are, in major part, bound!

3.3. Summing up the main points

The extensive discussion of the work of Granot et al. illustrates confusions which exist to this day in the field, and which interfere with the choice, the planning and the interpretation of relevant experiments. In an attempt to clarify matters in the future, we would like to discuss some key points in the argument.

1. Regarding Poisson-Boltzmann, the theory is well understood. Its shortcomings may be avoided by better theories such as cluster theory and others. However most experiments to date probably do not require such refinements. Rather, the model to which the theory is applied should be improved to include finite ion size, better description of the polyelectrolyte structure, binding interactions, etc.

2. Condensation theory is unsound. It should be challenged to provide the following :

a) A proper derivation for its predictions of counterion distributions.

b) A justification for rejecting site-binding.

c) An explanation of anomalous predictions, such as the persistence of condensation for a cylinder of vanishing surface charge density (a cylinder of increasing radius and constant linear charge density).

d) A clarification of its relation to Poisson-Boltzman theory.

e) A justification for the claim[23] that measured activity coefficients support the theory, when Poisson-Boltzmann is at least as successful in this respect.

f) An interpretation for experimental evidence demonstrating site-binding.

g) An interpretation for experimental evidence demonstrating that cylindrical geometry is not a key feature of polyelectrolytes.

3. Numerous definitions of binding have been given in the past, emphasizing either increased ionic concentration close to the polyelectrolyte or chemical interactions. One must first decide whether one is concerned with one or the other feature. One must then provide for a prescription to distinguish between them in case of ambiguity. For instance, in a solution of simple electrolytes, cations tend to cluster around anions. In dilute solutions of NaCl, it is usally considered that there is no chemistry involved. But

magnesium phosphate is different, since it will eventually precipitate. Of course, these problems may be studied by more refined methods which probe the structure and kinetics of the interactions, and the details of the distribution functions.

We think that an increasing body of opinion is in favor of defining binding as chemical interactions with groups of the polyelectrolyte. The groups may be charged or uncharged, and binding may be direct (inner sphere) or indirect (outer sphere). Binding can be detected by spectroscopic effects on counterions and ligands, such as broadening of phosphorus nmr by manganese. Such effects provide structural and kinetic information. A clearcut property of binding is that diffusion of counterions along and around the polyelectrolyte must be slowed down as compared to that expected for free ions.

Once agreement is reached on the definition of binding, it is still necessary to analyse the experiments. For instance the binding of sodium to DNA has given rise to conflicting interpretations which are directly traceable to the analysis of quadrupolar broadening.

ACKNOWLEDGMENTS

In preparing this text, I have drawn extensively on work performed in collaboration with Dr. J.L. Leroy and Prof. G. Weisbuch. N. Figueroa and P. Plateau contributed to the proton nmr of yeast $tRNA^{Asp}$, which was kindly provided by Dr. G. Keith.

REFERENCES

1. Rich, A. and Kim, S.H. (1978) Scientific American 238, 52-62

2. Cantor, C.R. and Schimmel P.R. (1980) "Biophysical Chemistry" (Freeman, San Francisco), Chapters 3,6,22-24

3. Leroy, J.L., Guéron, M., Thomas, G. and Favre, A.(1977) Eur. J. Biochem. 74, 567-574.

4. Shulman, R.G., Sternlicht, H. and Wyluda, B.J. (1965) J. Chem. Phys.,43, 3116-3122.

5. Leroy, J.L. and Guéron, M. (1981) Biochimie 63, 815-819

6. Guéron, M. and Leroy, J.L., in "ESR and NMR of Paramagnetic Species in Biological and Related Systems" (eds. I. Bertini and R.S. Drago ; Reidel, Dordrecht-Holland, 1979) 327-367.

7. Leroy, J.L. and Guéron, M. (1982) Biochimie 64, 297-299

8. Schimmel, P.R. and Redfield, A.G. (1980) Ann. Rev. Biophys. Bioeng. 9, 181-221.

9. Guéron, M. and Weisbuch, G. (1979) J. Phys. Chem. 83, 1991-1998

10. Guéron, M. and Weisbuch, G. (1980) Biopolymers, 19, 353-382

11.Morawetz, H. (1975) Macromolecules in Solution, 2nd ed., Wiley, New York, pp. 387-396

12. Fowler, R. and Guggenheim, E.A. (1960) Statistical Thermodynamics,Cambridge, Cambridge University Press, pp. 407-409

13. Iwasa, K. and Kwak, J.C.T. (1977) J. Phys. Chem. 81,408-412.

14. Schellman, J.A. and Stigter, D. (1977) Biopolymers 16, 1415-1434.

15. Leroy, J.L. and Guéron, M. (1977) Biopolymers 16,2429-2446.

16. Weisbuch, G. and Guéron, M. (1981) J. Phys. Chem. 85, 517-525.

17. Weisbuch, G. and Guéron, M., submitted to J. de Physique.

18. Gregor, M.P. and Gregor, J.M. (1977) J. Chem. Phys. 66 1934-1939

19. Jönsson, B., Wennerström, H. and Alle B., (1980) J. Phys. Chem. 84, 2179.

20. Grahame, D.C. (1947) Chem. Rev. 41, 441-501.

21. Fuoss, R.M., Katchalsky A. and Lifson S. (1951) Proc. Nat. Ac. Sci. US, 37, 579-589

22. Guéron,M. and Weisbuch, G. (1981) Biochimie 63, 815-819

23. Manning, G.S. (1978) Q. Rev. Biophys. 11, 179-246.

24. Stigter, D. (1978) J. Phys. Chem. 82, 1603-1606.

25. Anderson, C.F., Record, M.T., Jr. and Hart, P.A. (1978) Biophys. Chem. 7, 301-316.

26. (a) Granot, J., Feigon, J. and Kearns, D.R. (1982) Biopolymers 21, 181-201 ; (b) Granot, J. and Kearns, D.R., ibid, 203-218 and (c) 219-232

DIFFUSION OF CHARGED COLLOIDAL PARTICLES: SPATIAL AND DYNAMIC CORRELATIONS

Magdaleno Medina-Noyola
Departamento de Física
CINVESTAV-IPN, Apartado Postal 14-740
07000 México D.F. México

ABSTRACT

A simple theory is presented which predicts the main short- and long-time features observed in the intermediate scattering function F(k,t) measured in experiments of quasi elastic light scattering by suspensions of highly charged polystyrene macroparticles. This theory is based on the description of the coupled diffusion of the macroparticles and their counterions by the linear laws of irreversible thermodynamics, and on the use of spatially non-local thermodynamic relations which properly take into account the long-range correlations between the diffusing particles.

1. INTRODUCTION

During the last decade, light scattering techniques(1,2) have provided information on the static and dynamic properties of suspensions of interacting charged colloidal particles. This information is frequently expressed in terms of the intermediate scattering function, defined here as

$$F(k,t) \equiv \langle \delta n(\vec{k},t)\, \delta n(-\vec{k},t) \rangle / V \tag{1}$$

where $\delta n(\vec{k},t)$ is the spatial Fourier transform of the instantaneous fluctuation $\delta n(\vec{x},t)$ of the local macroparticle concentration around its average equilibrium value $n = N/V$, and the angular brackets reffer to an ensamble average. The Fourier transform in time of F(k,t) defines the dynamic structure factor $S(k,\omega)$, whereas the initial value $F(k,0) \equiv S(k)$ is the static structure factor. The theoretical interpretation of the experimental results is in general complicated since the explicit calculation of the intermediate scattering function requires the equivalent of a theory for the time evolution of the local macroparticle concentration $n(\vec{x},t)$, which depends on the dynamics of all the particles in the suspension. Under certain circumstances, however, simple phenomenological assumptions can be made which provide an adequate description of the main features of F(k,t) observed in specific experimental conditions. For example, for non-interacting Brownian particles, the quantity $\ln(F(k,t)/S(k))$ is found to be linear in k^2t, with a k-independent slope of magnitude equal to the diffusion constant D^o of the macroparticles. This experimental fact can be easily understood and derived(3) from the use of Fick's diffusion equation to describe the time evolution of $\delta n(\vec{x},t)$. For concentrated suspensions this simple picture is no longer valid, since many-body effects such as hydrodynamic interactions, and direct interaction forces between the macroparticles become important. In some cases, however, the direct interaction may be so long-ranged that even at very low concentrations, where hydrodynamic interactions can be neglected their effects on the structural properties are clearly apparent. This is the case in suspensions[1,2] of highly charged macroions at low ionic strength in which the static structure factor $S(k)$ exhibits liquid-like features that indicate that the range of the direct interaction is comparable to the average separation between the macroparticles. Under these circumstances, interesting features are also observed in the intermediate scattering function $F(k,t)$. Unlike the case of noninteracting Brownian particles, $F(k,t)$ is no longer a single exponential function of time. Still, an effective diffusion constant D_s can be defined[1,2] by the initial slope of $\ell nF((k,t)/S(k))$ plotted vs. k^2t, i.e.,

$$D_s(k) \equiv \lim_{t\to o} \{ - \frac{1}{k^2} \frac{\partial}{\partial t} \ell n(F(k,t)/S(k)) \} \tag{2}$$

This diffusion "constant", however, now becomes k-dependent, and the experimental results indicate that it is proportional to $S^{-1}(k)$. The overall time-dependence of $F(k,t)$ is not obvious from the experimental data, but a sum of two exponentials seems[2] to provide a good fit. For sufficiently long times, for examples, a single-exponential dependence is again apparent[2]. This behaviour can be characterized by a long-time, k-dependent diffusion "constant" $D_L(k)$ defined as

$$D_L(k) = \lim_{t\to\infty} \{ - \frac{1}{k^2} \frac{\partial}{\partial t} \ell n(F(k,t)/S(k)) \} \tag{3}$$

The k-dependence of $D_L(k)$ is observed[1,2] to imitate that of $D_s(k)$. For example, the peaks of $D_L^{-1}(k)$ occur at the same values of k as the maxima of $D_s^{-1}(k)$ (and of $S(k)$).

The number of recent works on the diffusion of interacting Brownian particles indicate a growing interest on the subject, mainly due to the availability of experimental information such as that mentioned above[1,2]. The theoretical work aimed at the interpretation of those experimental results starts from a variety of postulates (Generalized Smoluchowski equation[4-6], Langevin equation[7,8], etc) from which some of the properties of $F(k,t)$ can be derived using various theoretical techniques (projection operators[4], memory functions[5], Brownian dynamics[6], etc). Similarly, attempts have been made to explain the observed behaviour of $F(k,t)$ in terms of some sort of underlying mechanism ("cage" effects[1], direct interactions[7,8], plydispersity effects[9], etc). Some of these theories predict the observed k-dependence of $F(k,t)$ in the short-time regime, and an overall time dependence which is not a single exponential. However, no quantitatively satisfactory explanation is available of the k-dependence of the intermediate scattering function in the long-time regime, summarized by the long-time k-dependent diffusion constant of Eq. (3).

The theory described here[10] is also aimed at calculating F(k,t). The role of the static and dynamic coupling of the properties of the counterions with the charged colloidal particles, as the source of the observed behaviour of F(k,t) is investigated. The theory, which is explained in the following section, is based on a generalization of Fick's diffusion equation which properly takes into account the spatial correlations among the macroparticles and their counterions. The spatial correlations are calculated in section 3 from a well defined model of the interionic forces, and within a simple statistical mechanical approximation. These results are then used in section 4 to calculate the wave-vector dependence of F(k,t) in the long-time regime.

2. CALCULATION OF THE INTERMEDIATE SCATTERING FUNCTION.

Our theory starts from a phenomenological description of the diffusion of the macroparticles (species 1) and their counterions (species 2). It is assumed that the fluctuations $\delta n_i(\vec{x},t)$ of the local concentration of particles of species i obbey a stochastic version of the time-evolution equation that govern the decay of small but macroscopic departures of $\langle n_i(\vec{x},t)\rangle$ from their equilibrium value $n_i = N_i/V$, according to the linear laws of irreversible thermodynamics[3,11]. This is our basic dynamic postulate, and it leads to the following equations for $\delta n_i(\vec{x},t)$,

$$\frac{\partial \delta n_i(\vec{x},t)}{\partial t} = L_{ij}\, \nabla^2 \delta\beta\mu_j(\vec{x},t) + f_i(\vec{x},t) \tag{4}$$

Where i = 1,2 and with summation over repeated indices assumed. In this equation $T = 1/k_B\beta$ is the temperature, $\mu_j(\vec{x},t)$ is the local chemical potential of particles of species j, $L_{ij}(= L_{ji})$ is the matrix of phenomenological transport coefficients and $f_i(\vec{x},t)$ is a fluctuating "force". For a single component of noninteracting diffusing particles, eq. (4) is a stochastic version of Fick's diffusion equation.

A spatially non-local relation[10] is now assumed between the local chemical potentials and the local concentrations, so that the fluctuations $\delta\beta\mu_j(\vec{x},t)$ will be written as

$$\delta\beta\mu_j(\vec{x},t) = \int E_{jk}(|\vec{x}-\vec{x}'|)\,\delta n_k(\vec{x}',t)\,d\vec{x}' \tag{5}$$

where

$$E_{ij}(|\vec{x}-\vec{x}'|) \equiv \left(\frac{\delta\beta\mu_j[\vec{x};\underset{\sim}{n}]}{\delta n_k(\vec{x}')} \right) \tag{6}$$

is the functional derivative of $\beta\mu_j[\vec{x};n_1,n_2]$ with respect to $n_k(\vec{x}')$. This non-local thermodynamic derivative is related to the matrix of radial distribution functions $g_{ij}(r)$ by the Ornstein-Zernike equation, which can be written[12] in matrix form as

$$S(k)\hat{E}(k) = I \tag{7}$$

where $\hat{E}_{ij}(k)$ is the Fourier transform of $E_{ij}(r)$ and $S_{ij}(k)$ is the matrix of static structure factors, defined as

$$S_{ij}(k) \equiv n_i \{\delta_{ij} + n_j \int e^{-i\vec{k}\cdot\vec{r}} [g_{ij}(r) - 1] d\vec{r}\} \tag{8}$$

From the spatial Fourier transform of Eq. (4), using Eqs. (5)-(7), and assuming $<f_i(\vec{k},t)\delta n_k(-\vec{k},0)>$ to be zero, an equation can be derived for the matrix

$$F_{ij}(k,t) \equiv <\delta n_i(\vec{k},t)\delta n_j(-\vec{k},0)>/V \quad , \tag{9}$$

the (1,1) element of which is the intermediate scattering function defined in Eq. (1). This matrix is such that $F_{ij}(k,0) = S_{ij}(k)$, and its time-evolution is then determined by the following matrix equation

$$\frac{\partial F(k,t)}{\partial t} = - k^2 \, L \, S^{-1}(k) F(k,t) \tag{10}$$

The solution of this equation for the intermediate scattering function $F_{11}(k,t)$ is a sum of two exponentials,

$$\frac{F_{11}(k,t)}{S_{11}(k)} = \alpha_1(k) e^{-k^2 D_1(k) t} + \alpha_2(k) \, e^{-k^2 D_2(k) t} \tag{11}$$

where

$$\alpha_{1,2}(k) = [1 \pm a(k)] \, / \, 2, \tag{12}$$

$$D_{1,2}(k) = [c(k) \mp b(k)] \, / \, 2d(k), \tag{13}$$

$$a(k) = [c(k) - 2L_{11}/S_{11} d(k)] \, / \, b(k) \quad , \tag{14}$$

$$b(k) = [c^2(k) - 4 \frac{L_{11}}{S_{11}} \frac{L_{22}}{S_{22}} d(k) (1 - \frac{L_{12}^2}{L_{11} L_{22}})]^{1/2} \tag{15}$$

$$c(k) = \frac{L_{11}}{S_{11}} + \frac{L_{22}}{S_{22}} - \frac{2L_{12}}{S_{12}} (1 - d(k)) \tag{16}$$

and

$$d(k) = 1 - S_{12}^2 \, / \, S_{11} S_{22} \; . \tag{17}$$

These formulas indicate a rather involved dependence of $F_{11}(k,t)$ on the wave-vector, through the elements of $S_{ij}(k)$, and on the phenomenological coefficients L_{ij}. Thus, these matrices have to be determined before any explicit use of Eqs. (11)-(17). However, the short-time behaviour of $F_{11}(k,t)$, characterized by the initial slope of $\ln(F_{11}(k,t)/S_{11}(k))$, can be determined directly from those equations by calculating $D_s(k)$ defined in Eq. (2). The result reduces to the surprisingly simple and well known relation

$$D_s(k) = L_{11}/S_{11}(k) \tag{18}$$

This is a result which has found already full experimental verification[1,2]. At the same time, it provides in principle one way of experimentally determining the phenomenological coefficient L_{11} as the proportionality constant between $D_s(k)$ and $S^{-1}(k)$. From Eq. (18) it is clear that the dynamics of the counterions does not influence the short-time behaviour of $F_{11}(k,t)$, since only L_{11} appears in the result for $D_s(k)$. The transport coefficient L_{11} is related with the free particle diffusion constant of the macroparticles, D_1^o, and in the small concentration limit it is given by the Einstein-Stokes formula, $L_{11} = n_1 D_1^o = n_1/6\pi\beta\eta R_1$, where η is the viscosity of the solvent and R_1 is the diameter of the macroparticles. In this paper we will use this estimate of L_{11} and a similar one for L_{22} so that these parameters will be taken here as

$$L_{ii} = n_i/6\pi\beta\eta R_i \ , \quad i = 1,2 \tag{19}$$

No similar estimate can be made of the off-diagonal element $L_{12} = L_{21}$, which is a measure of the dynamic coupling between the diffusion of the macroparticles and the counterions. Thus, we decide to determine its value by the comparison of our results with the experimental data for $F_{11}(k,t)$. This procedure will be explained below. For the moment we turn our attention to the problem of calculations the matrix of static structure factors $S_{ij}(k)$.

3. THE STATIC STRUCTURE FACTOR

The calculation of the matrix $S_{ij}(k)$ is based[10] on the solution of the mean spherical closure of the Ornstein-Zernike equation,

$$h_{ij}(r) = c_{ij}(r) + \sum_{k=1}^{2} \rho_k \int h_{ik}(|\vec{r}-\vec{r}\,'|)c_{kj}(r')d\vec{r}\,' \tag{20}$$

In this equation, $h_{ij}(r) = g_{ij}(r)-1$ and $c_{ij}(r)$ is the direct correlation function. The interionic potentials are taken as those of the primitive model, i.e., hard-sphere plus coulombic interaction in a background of uniform dielectric constant ε (the solvent). Thus,

$$u_{ij}(r) = u^{H.S.}(r) + q_i\, q_j \,/\, \varepsilon r \tag{21}$$

where $u^{H.S.}(r) = 0$ for $r > R_{ij}$ and ∞ for $r < R_{ij} \equiv (R_i + R_j)/2$, and q_i and R_i are the charge and hard-sphere diameter of the particles of species i. For this model, the mean spherical closure is defined as $c_{ij}(r) = -\beta q_i q_j/\varepsilon r$ for $r > R_{ij}$ and $h_{ij}(r) = -1$ for $r < R_{ij}$. In the actual calculations presented here we used the fact that the ratios R_2/R_1 and q_2/q_1 for the system of interest are small ($<10^{-2}$). Thus, $S_{ij}(k)$ will be given by eq. (8), with $g_{ij}(r)$ obtained from the solution of the Ornstein-Zernike equation in the mean spherical closure in which the "dense point limit"[13] was taken.

In figs. 1 and 2 we compare the static structure factor $S_{11}(k)$ calculated in this manner with the experimental results of Grüner and Lehmann[2]. The values of the parameters used in the calculations are: $T = 298$ K, $\varepsilon = 78.5$, $R_1 = 900$ Å, $q_1 = 500\ e^-$ (e^- = electronic charge), and $n_1 = 2.53 \times 10^{12}$ cm^{-3} (fig. 1) and 1.265×10^{12} cm^{-3} (fig. 2). From these figures it is clear that the comparison, without being quantitatively fully satisfactory[10], it al least shows that the liquid-like features of $S_{11}(k)$ are qualitatively reproduced by our calculations. In any case it seems that for our present purpose, namely, the calculation of

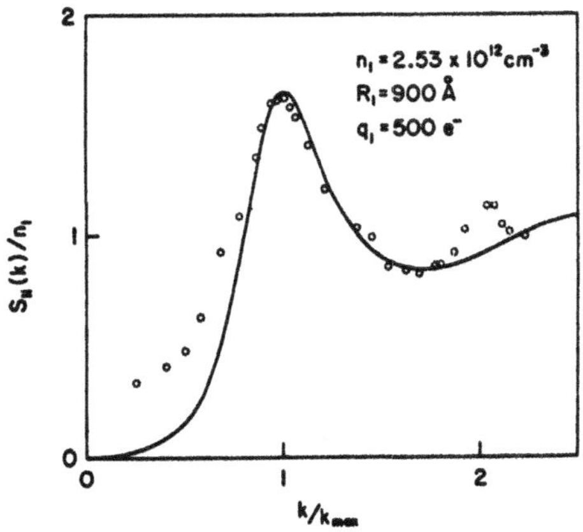

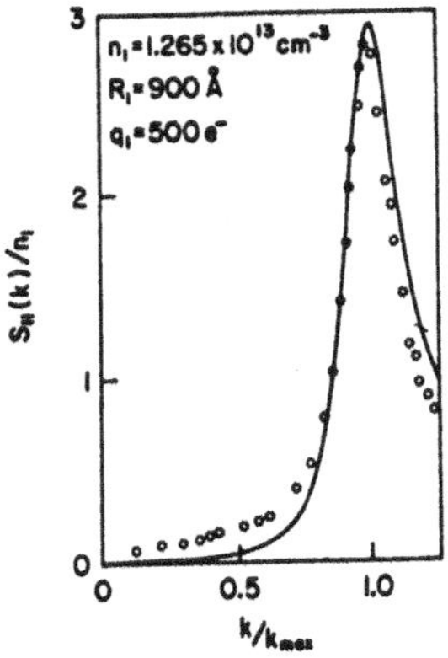

Figs. 1 + 2. Static structure factor $S_{11}(k)/n_1$ for the concentration indicated.

$F_{11}(k,t)$, the approximation described here for the static structure factors is entirely satisfactory.

4. THE LONG-TIME BEHAVIOUR OF F(k,t).

With the exception of the elements $L_{12} = L_{21}$, the matrix L_{ij} has been defined by our estimate in Eq. (19). In addition, $S_{ij}(k)$ were determined in the previous section. We can now calculate $F_{11}(k,t)$ according to Eqs. (11)-(17), using L_{12}, or, actually, $L_{12}^2/L_{11}L_{22}$ as an adjustable parameter. The long-time behaviour of $F_{11}(k,t)$ is better expressed in terms of $D_L(k)$ defined in Eq. (3). With a value of $L_{12}^2/L_{11}L_{22}$ of approximately 0.5, and with $q_2 = -e^-$ and an arbitrarily chosen value 4.5 Å for R_2, the theoretical results for $D_L^{-1}(k)$ are plotted in fig. 3 (solid line), along with $D_s^{-1}(k)$ (broken-dotted line). Both, D_L and D_s are scaled with $D_1^o = L_1/n = D_s(k\to\infty)$. This figure shows that another feature of the experimental data[2] is also present in the theoretical results. It consists on a k-dependence of $D_L(k)$ which clearly mimics that of $D_s(k) = L_{11}S_{11}^{-1}(k)$. Both, the experimental and our theoretical results in fig. 3 seem to indicate that the ratio $D_L(k)/D_s(k)$ has a simpler k-dependence than either D_s or D_L. According to our results, this ratio is in fact a constant except for $k < 0.5\, k_{max}$, where it departs from that constant value as $k \to 0$ in the manner indicated in fig. 3, where $[D_L(k)/D_s(k)]^{-1}$ has been also plotted (broken curve). This observation can now be used to establish a simple quantitative comparison between the theory and the experimental results. This is done in fig. 4, where the corresponding experimental data of Grüner and Lehmann[2] for $[D_L(k)/D_s(k)]^{-1}$ are compared with the theoretical results. We must mention that the constant value attained by $[D_L(k)/D_s(k)]$ was found[11] to be given by $[1 - L_{12}^2/L_{11}L_{22}]$, and this observation made the determination of $L_{12}^2/L_{11}L_{22}$ (= 0.47 in figs. (3) and (4)) simpler. From fig. 3 it is clear that the behaviour predicted by the theory is fairly well followed

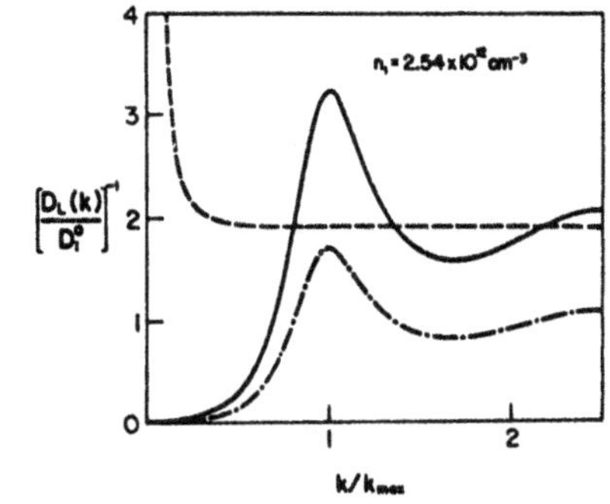

Fig. 3. $[D_L(k)/D_1^o]^{-1}$ (solid line), $[D_s(k)/D_1^o]^{-1}$ (broken-dotted line) and $[D_L(k)/D_s(k)]^{-1}$ (broken line).

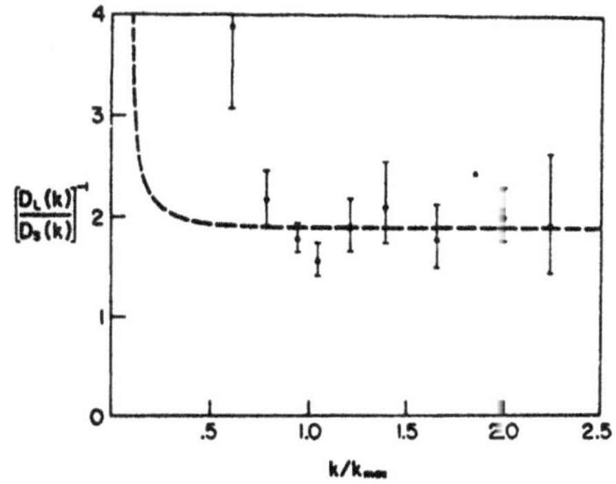

Fig. 4. Theoretical (broken line) and experimental results for $[D_L(k)/D_s(k)]^{-1}$.

by the experimental data. Further comparisons[10] indicate that the general trends observed in the available experimental results are in accord with those derived from the theory described here. Thus, accordance is found[10] in the increase of $[D_L(k)/D_s(k)]^{-1}$ with decreasing k and on the dependence on wave-vector and on concentration (at fixed k) of the amplitude $\alpha_1(k)$ defined in Eq. (12).

As a conclusion, we can say that comparisons of the type shown here indicate that not only the main features of $F_{11}(k,t)$ are qualitatively reproduced by our theory, but also that the quantitative agreement is remarkably good given the simplicity of the postulates and the rather primitive approximations made on the definition of the matrices L_{ij} and $S_{ij}(k)$ used in the specific calculations described here. Further quantitative comparisons, however, require refinements in our determination of L_{ij} and $S_{ij}(k)$ as well as more abundant and reliable experimental information on $F_{11}(k,t)$. The theoretical refinements, however, are naturally suggested by the methods employed here, and this is the subject of current work.

REFERENCES [1] P.N. Pusey, "Light Scattering in Fluids and Macromolecular Solutions", edited by V. Degiorgio, M. Corti and M. Giglio (plenum, N.Y., 1980). [2] F. Grüner and W. Lehmann. Ibid. [3] B. Berne, and R. Pecora. Dynamic Light scattering (Wiley, N.Y., 1976). [4] B. Ackerson. J. Chem. Phys. 69: 684 (1978). [5] W. Hess. "Light Scattering in Fluids and Macromolecular solutions", edited by V. Degiorgio, M. Corti and M. Giglio (Plenum, N.Y., 1980). [6] K. Gaylor, I. Snook and W. van Megen. J. Chem. Phys. 75: 1682 (1981). [7] P.N. Pursey and R.J.A. Tough. J. Phys. A15: 1291 (1982). [8] W. Hess and R. Klein. Physica 94A: 71 (1978). [9] M.B. Weissman. J. Chem. Phys. 72: 231 (1980). [10] M. Medina-Noyola. J. Chem. Phys. 77,1428,1434(1982). [11] S.R. de Groot and P. Mazur. "Nonequilibrium thermodynamics". (North Holland, Amsterdam, 1961). [12] M. Medina-Noyola and J. Keizer. Physica 107A: 437 (1981). [13] E. Waisman and J.L. Lebowitz, J. Chem. Phys. 56: 3093 (1972).

COUNTERION BINDING IN POLYELECTROLYTE THEORY+

J. Naghizadeh
Fachrichtung Theoretische Physik, Universität des Saarlandes, 66 Saarbrücken, FRG

The binding of counterions to a single highly charged polyion is formulated in terms of the two dimensional coulomb gas theory of Kosterlitz and Thouless. Mean field corrections to the binding criterion are calculated. These arise as a result of nonuniform counterion distribution in the vicinity of polyion.

It is well known that highly charged polyions whose linear charge density exceeds a certain treshold, bind with counterions present in their vicinity. The phenomenon was first described by Oosawa[1] and has been elaborated by Manning[2-4]. Counterion binding is important in biology as most biological molecules such as DNA and proteins are charged in solution. In what follows we limit our attention to an ideal linear polyion as distinct from the more complicated biological polyions (helical, double stranded helical, etc.). Because of the strong coulombic repulsion along the chain, the polyion is usually assumed to resemble a straight rod. This assumption has been experimentally justified.

Consider now a polyion of length L bearing P charged groups each with z_p elementary charges. Let $b=L/P$ be the average linear charge separation along the polyion. For simplicity, the polyion is assumed to be a uniformly charged rod of linear charge density γ

$$\gamma = \frac{q_p P}{L} = \frac{z_p e}{b} \tag{1}$$

where q_p is the charge of a single group on the polyion. The polyion is now immersed in a solvent medium of dielectric constant D and charged counterions of concentration n_1 per unit volume each of z_i elementary charges. The counterion binding criterion, established empirically and elaborated by Manning[2-4] is as follows. Let

$$\xi = e^2/Dbk_BT \tag{2}$$

with e the electronic charge k_B the Boltzmann constant and T the absolute temperature. For polyions of high charge density (small b) such that

$$\xi|z_i z_p| \geq 1 \tag{3}$$

sufficient counterions will condense on the polyion to bring

$$\xi|z_i z_p| = 1 \tag{4}$$

The binding is probably a consequence of the logarithmic potential between polyion and counterion[2,3]. In fact the condensation criterion (4) is quite similar to that arising in the problem of condensation of two dimesnional coulomb gas formulated by Kosterlitz and Thouless[5] and Hauge and Hemmer[6]. The condensation criterion (4) is only valid at high dilution. We wish to study its derivation in terms of the theory of Kosterlitz and Thouless and its possible modification due to screeing effects.

To proceed, the mutual potential U(r) of a polyion with uniform charge density and a counterion (for simplicity we take $z_i = z_p = 1$) in a salt solution of concentration n_1 counterions per unit volume may be arrived at approximately by the supersposition of the screened coulomb potential between the counterion and infinitesimal length of the polyion dz (taken along the polyion rod).With r the perpendicular distance form the rod, this is given by

$$U(r) = (e^2/bD) \int_{-\infty}^{+\infty} \frac{\exp -K(r^2+z^2)^{1/2}}{(r^2+z^2)^{1/2}} \, dz \tag{5}$$

$$U(r) = (2e^2/Db) \, K_o(Kr) \tag{6}$$

where $K_o(Kr)$ is a modified Bessel function of the second kind and

$$K^2 = \frac{4\pi e^2}{Dk_B T} (n_1 + n_2) \tag{7}$$

The quantity K is called the Debye screening parameter and n_1 and n_2 are the concentrations of the two species (positive and negative) of mobile ions present in the solution. The counterion concentration itself is composed of two types of ions (which may be the same) coming respectively from the dissociation of the polyion and the salt, whereas n_2 is exclusively due to salt. The potential U(r) takes a simple form at small distances and high dilution. In this limit ($Kr \rightarrow 0$)

$$K_o(kr) \approx -\ln(Kr) \tag{8}$$

At large distances, the function K_o decreases exponentially. Thus for $Kr >> 1$

$$K_o(Kr) \approx (\pi/2Kr)^{1/2} \exp(-Kr) \tag{9}$$

This behaviour ensures that, for the present problem where short distances are significant, the logarithmic potential of Eq. (8) may be taken to be valid everywhere, convergence of the integrals for large r being guaranteed by Eq. (9). In fact, the short range form of the potential is affected by discreet charge distribution on the polyion, and its structure (helical, etc.) These effects have been treated by Soumpasis[7]. They become significant at distances comparable to the average linear charge distance b. Although the condensation phenomenon occurs at these distances, the effect of discreetness is an additive term to the logarithmic potential, Eq. (8), and is not expected to alter the condensation criterion significantly.

Now following Kosterlitz and Thouless[5], we modify Eq. (8) in the following form

$$\begin{aligned} U(r_i) &= -(2e^2/Db)\ \ln(r_i/r_o) - (2e^2/Db)\ \ln(Kr_o) \quad & r_i > r_o \\ U(r_i) &= 0 & r_i < r_o \end{aligned} \tag{10}$$

where r_i is the distance of the i-th counterion from the polyion, and r_o is a distance of closest approach. This distance is introduced in order to avoid divergence of $U(r_i)$ at very short distances. Its justification lies in the fact that at very short distances repulsive forces between polyion and counterion become dominant. As seen from Eq. (10), $(2e^2/Db)\ \ln\ (Kr_o)$ represents the pair energy at distance r_o. We may now compute the average distance between counterion and polyion (ignoring for the present interactions between pairs)

$$\begin{aligned} \langle r^2 \rangle &= \frac{\int_{r_o}^{\infty} dr\ r^3 \exp\{-2\xi \ln(r/r_o)\}}{\int_{r_o}^{\infty} dr\ r\ \exp\{-2\xi \ln(r/r_o)\}} \\ &= r_o^2\, \frac{\xi-1}{\xi-2} \end{aligned} \tag{11}$$

The probability of finding a pair (counterion) at a distance r from polyion is given by

$$(Kr_o)^{-2\xi} \exp\{-2\ \ln(r/r_o)\} \tag{12}$$

From this, the average square distance between the pairs may be deduced as in coulomb gas problem

$$\begin{aligned} \langle 1/d^2 \rangle &= r_o^{-4}\ (Kr_o)^{-2\xi}(2\pi)\int_{r_o}^{\infty} r\ dr\ \exp\{-2\xi\ \ln(r/r_o)\} \\ &= r_o^{-2}\ \frac{(Kr_o)^{-2\xi}}{\xi - 2} \end{aligned} \tag{13}$$

Thus

$$\langle r^2/d^2 \rangle = \frac{(Kr_o)^{-2\xi}}{\xi - 2} \tag{14}$$

It is seen from Eq. (14) that the mean square distance between a counterion and the polyion diverges at

$$\xi = 2 \tag{15}$$

this is in contrast to the empirically estabished condensation criterion (4). The difference is a factor 2. The origin of this discrepancy may lie in the cutoff distance r_o. Hauge and Hemmer[6)] have found ξ=1 as the transition point without using a cutoff distance. It is also possible that the discrepancy is due to the structure of the polyion (helical, double helical etc.), as the coefficient of the logarithmic potential is sensitive to the structure as shown by Soumpasis[7)]. A further source of the discrepancy may be lie in the arbitrary definition of binding. Usually the ions within an arbitrary radius from the axis of the polyion are considered to be bound. In practice the counterion distribution in the vicinity of polyion is calculated by numercial solution of the Poisson-Boltzmann equation[9,8,10)]. This distribution is found to depend on the shape of the polyion. In fact the logarithmic potential is strictly valid for a polyion of zero cylinder-diameter. For finite diameters the conterion feels a charged curved surface, and shape factors become important. Some of these problems have been analysed by Gueron and Weisbuch[10)]. We shall defer further analysis of this problem and focus attention of the calculation of corrections to Eq. (15) arising from interactions between dipoles. This problem is entirely analogous to that of coulomb gas, if one uses instead of the pair potential (8) an effective pair potential given by

$$U_{eff}(r) \approx - (2e^2/DB) \ln \{(Kr^{\epsilon(r)})^{-1}\} \qquad (16)$$

where $\epsilon(r)$ is a slowly varying function arising from the mutual interactions between the counterions. It takes the value unity when these interactions are neglected. The physical significance of $\epsilon(r)$ is that, the counterions in the vicinity of the polyion are nonuniformly distributed. The consequences of the effective potential (16) have been worked out by Kosterlitz and Thouless. It leads, in the mean field approximations, to a modified condensation criterion given by

$$\xi = 2 \{1+1.3\pi\}(Kr_o)^{-\xi}\} \qquad (17)$$

The scheme presented here requires further analysis, in particular the origin of discrepancy between the calculated and empirical condensation criteria needs further clarification. However, the form of Eq. (16) indicates that most of studies based on the solution of the Poisson-Boltzmann equation may be erroneous in that they take a constant Debye screening parameter[8,9]. The screening parameter in the immediate vicinity of the polyion is indicated to be itself a function of distance.

Acknowledgement

The author is gratefull to Professors G.S. Manning and Arno Holz for stimulating discussions.

References

+ Work supported by Deutsche Forschungsgemeinschaft under SFB 130
1. F. Oosawa, J. Polymer Sci. 23, 421 (1957)
2. G.S. Manning, J. Chem. Phys. 51, 924 (1969)
3. G.S. Manning, J. Chem. Phys. 51, 3249 (1969)
4. G.S. Manning, Accounts Chem. Res. 12, 443 (1979)
5. J.M. Kosterlitz and D.J. Thouless, J. Phys. C6, 1181 (1973)
6. E. Hauge and P.C. Hemmer, Phys. Norvegica 5, 209 (1971)
7. D. Soumpasis, J. Chem. Phys. 69, 3190 (1978)
8. L. Kotin and M. Nagasawa, J. Chem. Phys. 36, 873 (1962)
9. A. Katchalsky, Z. Alexandrowics and O. Kedem in Chemical Physics of Ionic Solutions, B.E. Conway and R.G. Bradas Eds.,John Wiley, New York (1966)
10. M. Gueron and G. Weisbuch, Biopolymers 19, 353 (1980)

EFFECTIVE PROTEIN-PROTEIN POTENTIAL IN MEMBRANES

J. Naghizadeh

Fachrichtung Theoretische Physik, Universität des Saarlandes, 66 Saarbrücken, FRG

The effective lipid mediated protein-protein interaction has been calculated by Marcelja. He found that disruption of order of lipid molecules neighboring each protein is reduced as the number of lipids interposed between two proteins (protein-protein distance) is reduced. This gives rise to an attractive force between the proteins at short distances. The actual protein-protein interaction is deduced numerically from electron micrograms of lipids. These allow numerical computation of the radial distribution function g(R) of proteins in two dimensions. The g(R) is then inverted using methods of the theory of liquids to deduce the mutual potential U(R). The results are reasonable and agree with computations of Marcelja.

The effective lipid mediated protein interaction has recently been calculated by Marcelja[1]. In his calculation, the protein is taken to be rigid annulus which disturbs the order of its phospholipid environment leading to a change of free energy. The lipid mediated protein interaction is calculated by computing, in the mean field approximation, the change of free energy as a function of distance between two proteins. Marcelja's model predicts a decrease in the effective potential as the number of lipid chains interposed between the proteins decreases. The model does not take into account the bare protein-protein interaction.

It is possible to calculate the total protein-protein interaction with the aid of the electron microscope pictures of membrane surface. The protein distribution is visible in these photographs. The method involves numerical computation of the pair distribution of proteins g(R) defined by

$$g(R) = \frac{1}{n} \sum_{i=1}^{n} \frac{1}{2\pi\, R\varepsilon} \sum_{\substack{j \\ |\vec{r}_i - \vec{r}_j| - R < \varepsilon}} \qquad 1$$

Where n is the number of particles in a chosen region, and $\vec{r}_i$, $\vec{r}_j$ are the radius vectors of particles i and j respectively in an arbitrary two dimensional coordinate system defined on the electron microscope picture. The g(R) so defined is the mean number of particles in a shell or radius R and thickness ε around any chosen particle. It is known from the liquid theory that this pair distribution g(R) is related to the pair potential U(R). The exact form of the relationship is not known, but a number of approximate expressions relating g(R) and U(R) exist[2]. Experience gained from the liquids show that reasonable potential curves may be obtained with the approximate Hypernetted Chain (HNC) expression between g(R) and U(R)[2].

We have used the (HNC) approximation to calculate the total protein-protein potential with the aid of the numerically computed g(R). To do this, the following set of equations were solved numerically in two dimensions

$$h(R) = \frac{g(R)}{\langle n \rangle} - 1$$

$$c(R) = h(R) - \langle n \rangle \int c(|\vec{R} - \vec{R}'|)\, h(R')\, d\vec{R}'$$

$$\frac{U(R)}{kT} = h(R) - c(R) - \ln\,[h(R) + 1]$$

Here n is the mean particle density, k the Boltzmann constant and T the absolute temperature. The g(R) was calculated from the electron microscope photograph of the fracture face A for the plasma membrane of 3T3 Balb reported by da Silva and Martinez-Palomo[3]. Figure 1 shows the computed g(R) and Figure 2 the interaction potential as a function of the protein separation (the unit of length is the diameter of protein annulus). The results are insensitive of the choice of annulus diameters as long as this is in the repulsive region. The minimum in the potential occurs when the proteins are about 2,5 diameters apart. In addition there is a shoulder at about 1,5 diameter and a sharply repulsive part at low separations. In the absence of the lipid chains, one would expect a single minimum at about the location of the shoulder due to attractive forces between the two proteins. This expected minimum for bare proteins is obscured in the membrane due to the competition between protein-protein and lipid protein interactions. The minimum at 2,5 diameters most probably corresponds to the liquid mediated minimum of Marcelja.

The potential computed in this way, though approximate, is physically significant and in addition to reproducing the lipid mediated minimum, gives information about the interactions at low protein separations.

The location of the minimum is qualitatively consistent with the experimentally observed protein lattices in certain membranes.[4]

Thanks are due to Dr. S. Marcelja who suggested the problem leading to this work and to Miss U. Lippmann who helped in the computation of g(R). The author also wishes to thank professors Arno Holz and Martin Zuckermann for stimulating discussions. The financial support by NATO Scientific Affairs Division is gratefully acknowledged.

References

1. S. Marcelja, Biochem. Biophys. Acta 455, 1 (1976)
2. See for example P.A. Egelstaff, An Introduction ot the Liquid State (Academic Press, London, 1967)
3. P. Pinto da Silva and A. Martinez-Palomo, Proc. Nat. Acad. Sci. USA 72, 572 (1975)
4. W. Stoeckenius, in Structural Basis of Membrane Function, Y. Hatefi, L. Djavedi-Onaniance, Eds. (Academic Press, New York, 1976)

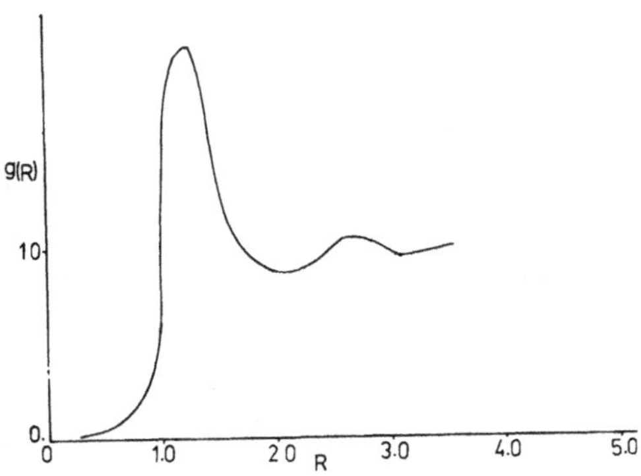

Fig. 1. Pair distribution function g(R) for proteins in the plasma membrane of 3T3 Balb (R in units of annulus diameter)

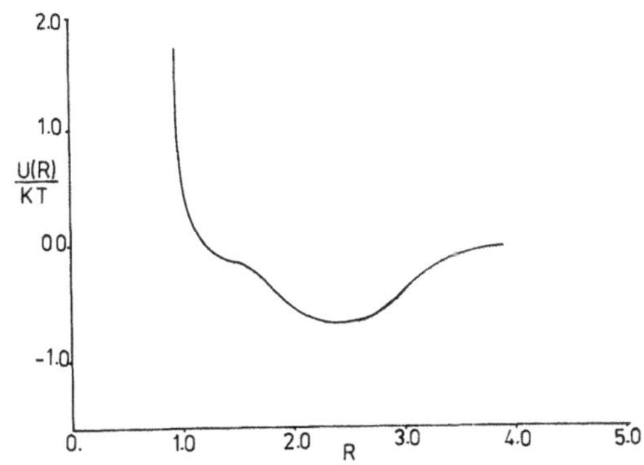

Fig. 2. u(R)/kT for proteins in the plasma membrane of 3T3 Balb (R in units of annulus diameter)

EQUILIBRIUM POLYMERIZATION OF STIFF POLYMERS

Marko V. Jarić and K.H. Bennemann
Freie Universität Berlin
Institute for Theoretical Physics
Arnimallee 14, D - 1000 Berlin 33

Abstract

Equilibrium polymerization of stiff polymers is considered in a mean-field approximation. It is shown that polymerization is of order-disorder type. The phase diagram exhibits a critical line which is replaced at tricritical point by a discontinuity line. It is also found that polymer weight fraction is non-zero at both sides of the transition showing only a mild singularity ($\sim|T-T_c|$) at continuous transition.

It is well known that polymer stiffness does not play an important role in statistics of long isolated polymers. This led to a belief that also in the case of equilibrium polymerization it suffices to treat a simpler case of flexible polymers. In recent papers Wheeler et al.[1] treated equilibrium polymerization of flexible polymers by going beyond chemical equilibrium theories[2]. They found that polymerization is a non-ordering transition with critical exponents of an n = o vector model. The transition occured at a single point, in the limit when energy for a polymer end went to infinity. This, however, appears to be inconsistent with a constraint, built into the model, that there are no polymer rings. Cordery[3] attempted to remove this inconsistency by allowing for ring formation but he had to give up excluded volume condition. Nevertheless, he found essentially same results except for critical exponents which he finds to be equal to the exponents of an n = 1 vector model.

Results for equilibrium polymerization of stiff polymers will be reported here. For reasons of simplicity the problem is solved on a square lattice in a mean-field approximation. An important feature of the model is that it unifies previous models[1,3] in treating both excluded volume and ring polymers. As a consequence dramatically different features of equilibrium polymerization are found. Although the model we use has some unphysical features we believe that it gives some insight into pecularities of equilibrium polymerization. Furthermore, there is a possibility that the mean-field results presented below can be considerably refined and improved.

Let us consider a square lattice[4]. At each site of the lattice we place either an inactivated monomer or an activated one. Activated monomer has two 'arms' (bifunctionality) which extend from the site half-way along (different) lattice bonds. Activation energy is denoted by $\varepsilon > 0$. Out of six configurations of activated monomers four are assigned an additional, bending, energy δ. Activated monomers also have 'sticky' ends leading to a release of an energy $v > 0$ whenever two arms (of two different monomers) meet at a lattice bond.

General formulation of this seven-state model with anisotropic nearest neighbour interactions is given elsewhere[4]. In the case $\delta > 0$ formation of straight polymers is preferred. The ground state is doublet (all polymers along $\hat{x}$ or along $\hat{y}$ directions) and polymerization is associated with symmetry breaking, i.e. ordering, in this doublet. Consequently, a mean-field theory is very simple. We denote by A_o concentration of unactivated monomers, by A_1 and A_3 concentrations of straight activated monomers in $\hat{x}$ and $\hat{y}$ directions, respectively, and by A_2 concentration of all other (bent) activated monomers. The variational free energy is[5]

$$\phi(A) = A_o \ln A_o + \sum_{i=1} A_i \ln A_i + A\cdot E - \tfrac{1}{2} A\cdot V\cdot A \tag{1}$$

where $A \equiv [A_1, A_2, A_3]$, $A_o \equiv 1 - \sum A_i$ and

$$E \equiv [\varepsilon, \varepsilon - \ln(H-2), \varepsilon] , \tag{2}$$

$$V \equiv v \begin{bmatrix} 2 & 1 & 0 \\ 1 & 1 & 1 \\ 0 & 1 & 2 \end{bmatrix} \tag{3}$$

We defined $H \equiv 2+4\exp(-\delta)$ which is a measure of the degrees of freedom per activated monomer. The free energy, Eq.(1), should be minimized with respect to A.

The results are exemplified by the limit of infinitely stiff polymers, $\delta \to \infty$ or $H \to 2$, which implies $A_2 \to 0$.[4,6]. We find that in this limit polymerization is always associated with symmetry breaking. This is best seen by considering weight fraction of polymers $Q = \sum A_i$ and the order parameter $M = A_1 - A_3$. In unpolymerized phase $M = 0$ while in polymerized phase $M \neq 0$. The critical line is given by

$$v_c = 1 + \frac{1}{2}\exp(\varepsilon_c - 1), \tag{4}$$

which ends at tricritical point

$$(\varepsilon_t, v_t) = (1 + \ln(4/3),\ 5/3), \tag{5}$$

where the transition becomes discontinuous. At the critical line M changes in a typical mean-field fashion, $\sim|T-T_c|^{1/2}$, while Q has milder singularity, $\sim|T-T_c|$.

In this model we can also easily calculate concentration C_{op} of (open) polymers

$$C_{op} = Q(1 - \frac{1}{2}Q) - \frac{1}{2}M^2, \tag{6}$$

which implies that average polymer length diverges only when system completely orders, i.e. when $M \rightarrow 1$ and $Q \rightarrow 1$. In the limit $\delta \rightarrow \infty$ we can also calculate concentration of polymers of any length l as

$$C_{lp} = A_1^l\ (1-A_1)^2 + A_3^l\ (1-A_3)^2, \tag{7}$$

which decays exponentially for large l. The average polymer length is

$$\langle l \rangle = Q/C_{op} \tag{8}$$

which can be obtained either directly or from Eq. (7).

Let us summarize the results of our treatment which are different in an essential way from the results obtained in other approaches[1,3]: (a) It is found here that polymerization is associated with an order-disorder transition, contrary to previous works[1,3]; (b) The transition between the polymer and non-polymer phases occurs along a line in (ε,v) plane (the transition is continuous below, and discontinuous above, tricritical point), whereas previous treatments find a transition at a single point ($v \rightarrow \infty$, $v - \varepsilon =$ const);[1,3] (c) Contrary to Ref. 1 we find that weight fraction of polymers is non-zero at both sides of the transition; (d) When the transition is continuous we find, similarly to Ref. 2, that Q shows only a mild singularity at the transition.

As already pointed out, the above conclusions remain also valid for finite δ, either positive or negative[4,6]. However, it is an open question whether a mean-field approach is indeed applicable to polymerization. The first reason, effects of fluctuations, although of general nature, is particularly emphasized here since for $\delta \leq 0$ the ground state is infinitely degenerated. Therefore, the phase diagram might be drastically altered even for $\delta > 0$. A hope that this will not happen comes from the fact that we included in our calculations all the ground states. Second objection to the mean-field treatment is that it effectively makes the interactions isotropic, whereas anisotropy seems to be essential to polymerization. An improvement in this direction might be sought, for example, within Oguchi's mean-field approach[7]. As an indication of the problems we also mention that for flexible polymers on hexagonal lattice exact results seem to disagree with the mean-field conclusions[8]. However, it is expected that for larger coordination numbers and in three dimensions the mean-field results are more realistic (note that in order to preserve ordering in three dimensions an additional interaction must be introduced[4,6]). It was also speculated recently[9] that polymerization might depend on the coordination number of the lattice, with, in particular, hexagonal lattice being pathological. Such lattice dependence, if proved correct, would seriously complicate interpretation of lattice models of polymerization and would question their applicability to real systems.

This work was supported in part by Sonderforschungsbereich 161. MVJ also acknowledges partial support through an Alexander von Humboldt-research fellowship.

References

1. J.C. Wheeler, S.J. Kennedy and P. Pfeuty, Phys. Rev. Lett. 45, 1748 (1980); J.C. Wheeler and P. Pfeuty, Phys. Rev. A23, 1531 (1981); J.C. Wheeler and P. Pfeuty, Phys. Rev. Lett. 46, 1409 (1981).
2. G. Gee, Trans. Faraday Soc. 48, 515 (1952); A.V. Tobolsky and A. Eisenberg, J. Am. Chem. Soc. 81, 780 (1959).
3. R. Cordery, Phys. Rev. Lett. 47, 457 (1981).
4. M.V. Jarić and K.H. Bennemann (to be published).
5. H. Falk, Am. J. Phys. 38, 858 (1970).
6. A detailed treatment for both $\delta > 0$ and $\delta \leq 0$ (ground state infinitely degenerated) will be presented elsewhere .
7. S. Smart. Effective Field Theories of Magnetism (W.B. Saurders Company, Philadelphia 1966).
8. M.V. Jarić and B. Nienhuis (in preparation)
9. F. Rys, private communications.

Lecture Notes in Physics

Vol. 136: The Role of Coherent Structures in Modelling Turbulence and Mixing. Proceedings 1980. Edited by J. Jimenez. XIII, 393 pages. 1981.

Vol. 137: From Collective States to Quarks in Nuclei. Edited by H. Arenhövel and A. M. Saruis. VII, 414 pages. 1981.

Vol. 138: The Many-Body Problem. Proceedings 1980. Edited by R. Guardiola and J. Ros. V, 374 pages. 1981.

Vol. 139: H. D. Doebner, Differential Geometric Methods in Mathematical Physics. Proceedings 1981. VII, 329 pages. 1981.

Vol. 140: P. Kramer, M. Saraceno, Geometry of the Time-Dependent Variational Principle in Quantum Mechanics. IV, 98 pages. 1981.

Vol. 141: Seventh International Conference on Numerical Methods in Fluid Dynamics. Proceedings. Edited by W. C. Reynolds and R. W. MacCormack. VIII, 485 pages. 1981.

Vol. 142: Recent Progress in Many-Body Theories. Proceedings. Edited by J. G. Zabolitzky, M. de Llano, M. Fortes and J. W. Clark. VIII, 479 pages. 1981.

Vol. 143: Present Status and Aims of Quantum Electrodynamics. Proceedings, 1980. Edited by G. Gräff, E. Klempt and G. Werth. VI, 302 pages. 1981.

Vol. 144: Topics in Nuclear Physics I. A Comprehensive Review of Recent Developments. Edited by T.T.S. Kuo and S.S.M. Wong. XX, 567 pages. 1981.

Vol. 145: Topics in Nuclear Physics II. A Comprehensive Review of Recent Developments. Proceedings 1980/81. Edited by T. T. S. Kuo and S. S. M. Wong. VIII, 571-1.082 pages. 1981.

Vol. 146: B. J. West, On the Simpler Aspects of Nonlinear Fluctuating. Deep Gravity Waves. VI, 341 pages. 1981.

Vol. 147: J. Messer, Temperature Dependent Thomas-Fermi Theory. IX, 131 pages. 1981.

Vol. 148: Advances in Fluid Mechanics. Proceedings, 1980. Edited by E. Krause. VII, 361 pages. 1981.

Vol. 149: Disordered Systems and Localization. Proceedings, 1981. Edited by C. Castellani, C. Castro, and L. Peliti. XII, 308 pages. 1981.

Vol. 150: N. Straumann, Allgemeine Relativitätstheorie und relativistische Astrophysik. VII, 418 Seiten. 1981.

Vol. 151: Integrable Quantum Field Theory. Proceedings, 1981. Edited by J. Hietarinta and C. Montonen. V, 251 pages. 1982.

Vol. 152: Physics of Narrow Gap Semiconductors. Proceedings, 1981. Edited by E. Gornik, H. Heinrich and L. Palmetshofer. XIII, 485 pages. 1982.

Vol. 153: Mathematical Problems in Theoretical Physics. Proceedings, 1981. Edited by R. Schrader, R. Seiler, and D.A. Uhlenbrock. XII, 429 pages. 1982.

Vol. 154: Macroscopic Properties of Disordered Media. Proceedings, 1981. Edited by R. Burridge, S. Childress, and G. Papanicolaou. VII, 307 pages. 1982.

Vol. 155: Quantum Optics. Proceedings, 1981. Edited by C.A. Engelbrecht. VIII, 329 pages. 1982.

Vol. 156: Resonances in Heavy Ion Reactions. Proceedings, 1981. Edited by K.A. Eberhard. XII, 448 pages. 1982.

Vol. 157: P. Niyogi, Integral Equation Method in Transonic Flow. XI, 189 pages. 1982.

Vol. 158: Dynamics of Nuclear Fission and Related Collective Phenomena. Proceedings, 1981. Edited by P. David, T. Mayer-Kuckuk, and A. van der Woude. X, 462 pages. 1982.

Vol. 159: E. Seiler, Gauge Theories as a Problem of Constructive Quantum Field Theory and Statistical Mechanics. V, 192 pages. 1982.

Vol. 160: Unified Theories of Elementary Particles. Critical Assessment and Prospects. Proceedings, 1981. Edited by P. Breitenlohner and H.P. Dürr. VI, 217 pages. 1982.

Vol. 161: Interacting Bosons in Nuclei. Proceedings, 1981. Edited by J.S. Dehesa, J.M.G. Gomez, and J. Ros. V, 209 pages. 1982.

Vol. 162: Relativistic Action at a Distance: Classical and Quantum Aspects. Proceedings, 1981. Edited by J. Llosa. X, 263 pages. 1982.

Vol. 163: J. S. Darrozes, C. Francois, Mécanique des Fluides Incompressibles. XIX, 459 pages. 1982.

Vol. 164: Stability of Thermodynamic Systems. Proceedings, 1981. Edited by J. Casas-Vázquez and G. Lebon. VII, 321 pages. 1982.

Vol. 165: N. Mukunda, H. van Dam, L.C. Biedenharn, Relativistic Models of Extended Hadrons Obeying a Mass-Spin Trajectory Constraint. Edited by A. Böhm and J.D. Dollard. VI, 163 pages. 1982.

Vol. 166: Computer Simulation of Solids. Edited by C.R.A. Catlow and W.C. Mackrodt. XII, 320 pages. 1982.

Vol. 167: G. Fieck, Symmetry of Polycentric Systems. VI, 137 pages, 1982.

Vol. 168: Heavy-Ion Collisions. Proceedings, 1982. Edited by G. Madurga and M. Lozano. VI, 429 pages. 1982.

Vol. 169: K. Sundermeyer, Constrained Dynamics. IV, 318 pages. 1982.

Vol. 170: Eighth International Conference on Numerical Methods in Fluid Dynamics. Proceedings, 1982. Edited by E. Krause. X, 569 pages. 1982.

Vol. 171: Time-Dependent Hartree-Fock and Beyond. Proceedings, 1982. Edited by K. Goeke and P.-G. Reinhard. VIII, 426 pages. 1982.

Vol. 172: Ionic Liquids, Molten Salts and Polyelectrolytes. Proceedings, 1982. Edited by K.-H. Bennemann, F. Brouers, and D. Quitmann. VII, 253 pages. 1982.

Selected Issues from

Lecture Notes in Mathematics

Vol. 809: I. Gumowski and Ch. Mira, Recurrences and Discrete Dynamic Systems. VI, 272 pages. 1980.

Vol. 810: Geometrical Approaches to Differential Equations. Proceedings 1979. Edited by R. Martini. VII, 339 pages. 1980.

Vol. 816: L. Stoica, Local Operators and Markov Processes. VIII, 104 pages. 1980.

Vol. 819: Global Theory of Dynamical Systems. Proceedings, 1979. Edited by Z. Nitecki and C. Robinson. IX, 499 pages. 1980.

Vol. 821: Statistique non Paramétrique Asymptotique. Proceedings, 1979. Edited by J.-P. Raoult. VII, 175 pages. 1980.

Vol. 826: Ph. G. Ciarlet et P. Rabier, Les Equations de von Kármán. VI, 181 pages. 1980.

Vol. 827: Ordinary and Partial Differential Equations. Proceedings, 1978. Edited by W. N. Everitt. XVI, 271 pages. 1980.

Vol. 831: Representation Theory I. Proceedings, 1979. Edited by V. Dlab and P. Gabriel. XIV, 373 pages. 1980.

Vol. 832: Representation Theory II. Proceedings, 1979. Edited by V. Dlab and P. Gabriel. XIV, 673 pages. 1980.

Vol. 836: Differential Geometrical Methods in Mathematical Physics. Proceedings, 1979. Edited by P. L. García, A. Pérez-Rendón, and J. M. Souriau. XII, 538 pages. 1980.

Vol. 837: J. Meixner, F. W. Schäfke and G. Wolf, Mathieu Functions and Spheroidal Functions and their Mathematical Foundations Further Studies. VII, 126 pages. 1980.

Vol. 838: Global Differential Geometry and Global Analysis. Proceedings 1979. Edited by D. Ferus et al. XI, 299 pages. 1981.

Vol. 840: D. Henry, Geometric Theory of Semilinear Parabolic Equations. IV, 348 pages. 1981.

Vol. 841: A. Haraux, Nonlinear Evolution Equations- Global Behaviour of Solutions. XII, 313 pages. 1981.

Vol. 842: Séminaire Bourbaki vol. 1979/80. Exposés 543–560. IV, 317 pages. 1981.

Vol. 843: Functional Analysis, Holomorphy, and Approximation Theory. Proceedings. Edited by S. Machado. VI, 636 pages. 1981.

Vol. 845: A. Tannenbaum, Invariance and System Theory: Algebraic and Geometric Aspects. X, 161 pages. 1981.

Vol. 849: P. Major, Multiple Wiener-Itô Integrals. VII, 127 pages. 1981.

Vol. 851: Stochastic Integrals. Proceedings, 1980. Edited by D. Williams. IX, 540 pages. 1981.

Vol. 852: L. Schwartz, Geometry and Probability in Banach Spaces. X, 101 pages. 1981.

Vol. 856: R. Lascar, Propagation des Singularités des Solutions d'Equations Pseudo-Différentielles à Caractéristiques de Multiplicités Variables. VIII, 237 pages. 1981.

Vol. 858: E. A. Coddington, H. S. V. de Snoo: Regular Boundary Value Problems Associated with Pairs of Ordinary Differential Expressions. V, 225 pages. 1981.

Vol. 861: Analytical Methods in Probability Theory. Proceedings 1980. Edited by D. Dugué, E. Lukacs, V. K. Rohatgi. X, 183 pages. 1981.

Vol. 866: J.-M. Bismut, Mécanique Aléatoire. XVI, 563 pages. 1981.

Vol. 878: Numerical Solution of Nonlinear Equations. Proceedings, 1980. Edited by E. L. Allgower, K. Glashoff, and H.-O. Peitgen. XIV, 440 pages. 1981.

Vol. 881: R. Lutz, M. Goze, Nonstandard Analysis. XIV, 261 pages. 1981.

Vol. 888: Padé Approximation and its Applications. Proceedings, 1980. Edited by M. G. de Bruin and H. van Rossum. VI, 383 pages. 1981.

Vol. 898: Dynamical Systems and Turbulence, Warwick, 1980. Proceedings. Edited by D. Rand and L.-S. Young. VI, 390 pages. 1981.

Vol. 901: Séminaire Bourbaki vol. 1980/81 Exposés 561–578 III, 299 pages. 1981.

Vol. 904: K. Donner, Extension of Positive Operators and Korovkin Theorems. XII, 182 pages. 1982.

Vol. 905: Differential Geometric Methods in Mathematical Physics. Proceedings, 1980. Edited by S.J. Andersson, H.-D. Doebner, and H.R. Petry. VI, 309 pages. 1982.

Vol. 909: Numerical Analysis. Proceedings, 1981. Edited by J.P. Hennart. VII, 247 pages. 1982.

Vol. 912: Numerical Analysis. Proceedings, 1981. Edited by G. A. Watson. XIII, 245 pages. 1982.

Vol. 920: Séminaire de Probabilités XVI, 1980/81. Proceedings. Edité par J. Azéma et M. Yor. V, 622 pages. 1982.

Vol. 921: Séminaire de Probabilités XVI, 1980–8 Supplément: Géométrie Différentielle Stochastique. Proceedings Edité par J. Azéma et M. Yor. III, 285 pages. 1982.

Vol. 922: B. Dacorogna, Weak Continuity and Weak Lower Semicontinuity of Non-Linear Functionals. V, 120 pages. 1982.

Vol. 923: Functional Analysis in Markov Processes. Proceedings, 1981. Edited by M. Fukushima. V, 307 pages. 1982.

Vol. 926: Geometric Techniques in Gauge Theories. Proceedings, 1981. Edited by R. Martini and E.M.de Jager. IX 219 pages. 1982.

Vol. 927: Y. Z. Flicker, The Trace Formula and Base Change for GL (3). XII, 204 pages. 1982.

Vol. 928: Probability Measures on Groups. Proceedings 1981. Edited by H. Heyer. X, 477 pages. 1982.

Vol. 929: Ecole d'Eté de Probabilités de Saint-Flour X – 1980. Proceedings, 1980. Edited by P.L. Hennequin. X, 313 pages. 1982.

Vol. 930: P. Berthelot, L. Breen, et W. Messing, Théorie de Dieudonné Cristalline II. XI, 261 pages. 1982.

Vol. 931: D.M. Arnold, Finite Rank Torsion Free Abelian Groups and Rings. VII, 191 pages. 1982.

Vol. 932: Analytic Theory of Continued Fractions. Proceedings, 1981. Edited by W.B. Jones, W.J. Thron, and H. Waadeland. VI, 240 pages. 1982.

Vol. 934: M. Sakai, Quadrature Domains. IV, 133 pages. 1982.

Vol. 935: R. Sot, Simple Morphisms in Algebraic Geometry. IV, 146 pages. 1982.

Vol. 936: S.M. Khaleelulla, Counterexamples in Topological Vector Spaces. XXI, 179 pages. 1982.

Vol. 937: E. Combet, Intégrales Exponentielles. VIII, 114 pages. 1982.

Vol. 938: Number Theory. Proceedings, 1981. Edited by K. Alladi. IX, 177 pages. 1982.

Vol. 942: Theory and Applications of Singular Perturbations. Proceedings, 1981. Edited by W. Eckhaus and E.M. de Jager. V, 363 pages. 1982.

Vol. 953: Iterative Solution of Nonlinear Systems of Equations. Proceedings, 1982. Edited by R. Ansorge, Th. Meis, and W. Törnig. VII, 202 pages. 1982.

GPSR Compliance

The European Union's (EU) General Product Safety Regulation (GPSR) is a set of rules that requires consumer products to be safe and our obligations to ensure this.

If you have any concerns about our products, you can contact us on ProductSafety@springernature.com

In case Publisher is established outside the EU, the EU authorized representative is:

Springer Nature Customer Service Center GmbH
Europaplatz 3
69115 Heidelberg, Germany

Batch number: 10704763

Printed by Printforce, the Netherlands